RACING GREEN

Also available in the Bloomsbury Sigma series:

Sex on Earth by Jules Howard
Spirals in Time by Helen Scales
A is for Arsenic by Kathryn Harkup
Suspicious Minds by Rob Brotherton
Herding Hemingway's Cats by Kat Arney
The Tyrannosaur Chronicles by David Hone
Soccermatics by David Sumpter
Goldilocks and the Water Bears by Louisa Preston
Science and the City by Laurie Winkless
Built on Bones by Brenna Hassett
The Planet Factory by Elizabeth Tasker
Catching Stardust by Natalie Starkey
Nodding Off by Alice Gregory
Turned On by Kate Devlin
Borrowed Time by Sue Armstrong
The Vinyl Frontier by Jonathan Scott
Clearing the Air by Tim Smedley
Superheavy by Kit Chapman
The Contact Paradox by Keith Cooper
Life Changing by Helen Pilcher
Kindred by Rebecca Wragg Sykes
Our Only Home by His Holiness The Dalai Lama
First Light by Emma Chapman
Ouch! by Margee Kerr & Linda Rodriguez McRobbie
Models of the Mind by Grace Lindsay
The Brilliant Abyss by Helen Scales
Overloaded by Ginny Smith
Handmade by Anna Ploszajski
Beasts Before Us by Elsa Panciroli
Our Biggest Experiment by Alice Bell
Worlds in Shadow by Patrick Nunn
Aesop's Animals by Jo Wimpenny
Fire and Ice by Natalie Starkey
Sticky by Laurie Winkless

RACING GREEN

HOW MOTORSPORT SCIENCE CAN SAVE THE WORLD

Kit Chapman

BLOOMSBURY SIGMA
LONDON • OXFORD • NEW YORK • NEW DELHI • SYDNEY

BLOOMSBURY SIGMA
Bloomsbury Publishing Plc
50 Bedford Square, London, WC1B 3DP, UK
29 Earlsfort Terrace, Dublin 2, Ireland

First published in the United Kingdom in 2022

Photo credits (t = top, b = bottom, l = left, r = right, c = centre)

Colour section: P. 1: © Public domain (t); © Venturi (b). P. 2: © Kit Chapman (t); © Public domain, Bain News Service (b). P. 3: © FIA Formula E (t); © Kit Chapman (c, b). P. 4: © McLaren (t); ©Hexashots, Olivier de Groot (b). P. 5: © Wirth Research (t, c, b). P. 6: © James Tye / UCL (t); © UCL (c); © James Tye / UCL (b). P. 7: © Public domain (tl); Wellcome Collection (tr); © Dale Rodgers (b). P. 8: © Roborace (t, b).

A catalogue record for this book is available from the British Library

Library of Congress Cataloguing-in-Publication data has been applied for

ISBN: HB: 978-1-4729-8217-9; TPB: 978-1-4729-8216-2; eBook: 978-1-4729-8218-6;

2 4 6 8 10 9 7 5 3 1

Typeset by Deanta Global Publishing Services, Chennai, India
Printed and bound in Great Britain by CPI Group (UK) Ltd, Croydon CR0 4YY
Bloomsbury Sigma, Book Seventy-one

Contents

Research for this book started in late 2019 and continued into 2021. As such, the Covid-19 pandemic disrupted much of its planned content. Visits to Williams, Silverstone and more were scrapped; face-to-face chats were moved online; races I was planning to attend never took place.

At the end of the day, though, this is just a book. The Covid-19 pandemic cost lives. This work is dedicated to all those who will never see live sport again, including my uncle, Stuart Chambers.

Introduction

Romain Grosjean has 27 seconds to live.

It's Sunday, 29 November 2020. The Bahrain Grand Prix has just started, the roar and sparks of the hybrid engines creating thunder and lightning in the Gulf night. Twenty of the most advanced cars on the planet jostle for position. Through turn three, Grosjean darts to the right to avoid a wall of slower opponents in his path. He arrows for clear air, attempting to thread his car at 150mph through an ever-narrowing gap between Haas teammate Kevin Magnussen and the AlphaTauri of Daniil Kvyat.

It's a gap that doesn't exist.

Grosjean's rear right tyre clips Kvyat's front left, sending him skidding off track in a shower of burnt rubber. He smashes into the safety barrier, cutting through a metal wall with jagged edges that slice his car in half. In a thousandth of a second, high-octane fuel spills, ignites and explodes in a spectacular fireball. The car vanishes from sight, lost in a sheet of burning death. Grosjean is trapped in a broken wreck at the heart of the inferno.

My heart thuds, seems to catch for a moment, pulses again. My breath judders out a gasp of terror. My eyes blink, wondering if what I've just seen is real. It shouldn't have been possible for a car to ignite like that. Is he alright? Is he *alive*? The fearsome, terrible orange flame continues to soar higher into the clear desert skies. The cars in the race continue, no driver oblivious to the carnage, but all knowing that to stop would only lead to further accidents and potential loss of life. In the background, as the camera pans away, the medical car can be seen rushing to Grosjean's aid. It's at the scene in seconds, along with fire marshals and extinguishers, to attempt a desperate rescue.

I have nothing to do with the drama unfolding. I'm half a world away, skulking in a South Korean hostel with the faint smell of spiced meats drifting up from the kitchen below. And yet the images on the TV suck me into a vortex to relive an indelible, haunting memory. I'm a child again, waking up on the morning of 2 May 1994, asking my mum why she's crying as she makes breakfast. That night, she had been watching the San Marino Grand Prix. Ayrton Senna – the brilliant, passionate, aggressive name among names – had swerved off track and crashed hard into a concrete barrier. He had died with the world watching.

I was too young and stupid to process such utter calamity. Senna was the baddie, the man who had so often beaten my hero, Nigel Mansell, the guy who drove a McLaren slathered with Marlboro slogans that looked like a giant, crushed pack of cigarettes. I couldn't comprehend the prodigious talent that had once seen him win a Grand Prix despite being stuck in sixth gear. I didn't see how his driving lit up and graced the world with a deft, balletic control that left his rivals awestruck as he slipped past. I had no notion of the raw, senseless waste that had led to a man dying for my entertainment. I just nodded and went to school, where the news finally sank into my soul. It was as if someone had reached out with an invisible hand and snatched all colour from the universe, transforming it into faded and muffled monochrome. My life until then had been blessed with little tragedy; Senna's death was the day I grew up.

I feel the same muted palette seeping into my vision as I watch Grosjean die. But the colours won't fade and the flames lose none of their lustre as they shroud his broken coffin. I hope for life. I expect less. I saw the crash that caused the death of Jules Bianchi, 21 years after Senna's. I saw the loss of Anthoine Hubert at Spa in 2019. But Grosjean's crash hits harder because, moments before, I was writing about the technology that could save him. Of halos and quick releases, of flame-retardant suits and the changes made by medics and marshals. I sang of heroic science and the science of heroes. And I now don't know if it will be enough.

Wait. Something's moving within the glow. Please, please, please…

YEEEEEEEEEEEEEEESSSSSSSSSSSSSSSSSS!

A figure emerges from the flames. Wading through fire, scrambling across warped metal with the help of first responders, Romain Grosjean escapes the roaring furnace. His car is a charred husk. He has lost a shoe. He pulls his gloves away to reveal burnt hands licked by Hell on Earth, but he is walking and breathing and talking. The world and I scream in relief.

For 27 seconds, Romain Grosjean should have died. That he survived with minor wounds – burnt hands and an injured foot – is a testament to the life-saving power of science.

Suddenly, writing this book takes on a new purpose. This is a story of invention, of myriad discoveries, ideas and technologies developed through racing and how they have an astonishing, hidden impact on our lives. It covers bold thinking, creative solutions and green advances that will help curb the impending doom of climate catastrophe.

But it's also about something far simpler than any of that. This is a book is about how racing cars will save your life. It is a glimpse into the hidden boons of motorsport – the weird, unlikely ways that efforts to gain a sliver more time on the track have also given us a sliver more time on Earth.

Motorsport is often dismissed as a trivial, environmentally harmful, perilous spectacle. For critics, it's a modern chariot race, horses and whips replaced by petrol-guzzling cars that spew out noise and fury for the joy of millions. As with their ancient counterparts, the racers battle at the edge of human endurance and capability. They drive custom creations honed by experts for speed and handling. They gamble their lives for that extra half-second between glory and defeat. They even get laurel wreaths to signify victory. At the end of the day, it's just a bunch of cars going around in circles, right?

Not even close. It's so much more. I don't see chariots racing around the Circus Maximus as the mob bays for blood. I see the world's fastest R&D lab. It's a place where we are reminded that the word 'engineer' doesn't come from someone who maintains engines; it's from the Latin *ingenium*, meaning 'cleverness'.

Elite sport is always an arms race, a constant battle in pursuit of excellence that requires a team of hundreds to stay competitive. Usually, however, science is there to support the talent of its competitors. In the world of motor racing, it's the other way around. Every ounce of a Formula One car is weighed, measured and accounted for; every wing and curve is a design choice; every groove on the tyre the result of countless experiments and experiences. While Sir Lewis Hamilton and Max Verstappen might get the plaudits (and both deserve all they receive and more), the truth is they are just the meat in the machine, bound by the limits of the cars they drive.

These design choices, made with the sole aim of making a car go a second, tenth of a second or hundredth of a second faster than a competitor, are small miracles. If they work – and sometimes even if they don't – they invariably end up rippling through our homes and communities. It's the astonishing aerodynamics of a Formula One car that keeps your lunch cold in a supermarket fridge. It's Formula E, a new kid on the block using city streets and fully electric single-seater racers, that saw electric cars go from an eccentric folly to the undisputed future of the automotive industry. And, if you trace the origins of simulated humans, computer programmes that can create a virtual *you* to help doctors treat disease, you'll find they begin with a brilliant mathematician arguing with David Coulthard about his Xbox.

Motorsport has helped transform safety – not only on the track and on our roads, but much farther afield. Today, a child could be rushed to Great Ormond Street Hospital in London in a protective cradle built by Williams, receive care from a surgical team modelled after a Ferrari pit crew and be monitored on the ward with a device from McLaren. As unlikely as it seems, the death of Senna played an invisible role in the advent of autonomous vehicles, helping them to communicate with each other as they navigate our increasingly busy streets. When the Covid-19 pandemic struck, it was the astonishing engineering prowess of the Mercedes-AMG Formula One team that helped build 10,000 breathing aids in less than a month and saved London's hospitals from being completely overwhelmed.

But this is only the beginning. Climate change has put the world on a precipice and new innovations are needed if future generations are to thrive. New materials, methods of production and means of producing energy must emerge within the next decade. It's here that motorsport excels too – as a testbed for ideas and principles we will need to adopt. Already, you can see this in action, whether it's in racing cars with flax bodies and dandelion tyres, a 3D-printed Shelby Cobra or the student teams helping to miniaturise hydrogen fuel cells and make them a viable alternative.

Motorsport has already changed the world.

Its next challenge is to save it.

PART ONE

THE FASTEST R&D LAB ON EARTH

CHAPTER ONE

Fire and ICE

Camille Jenatzy sat inside a torpedo. At least, that's how it appeared to the crowd gathered along an arrow-straight lane outside Achères, a few miles north-west of Paris, France. It was a little after 3 p.m. on 29 April 1899 and the grey, cigar-shaped body of his craft glistened from the light rain that had fallen all day. The machine was unlike anything the world had seen before – a smooth, aerodynamic bullet above a simple chassis attached to four wheels with thick, wide tyres. The name of Jenatzy's engineering aberration was *La Jamais Contente* ('The Never Satisfied'). It was the world's first purpose-built racing car, created to shatter the land-speed record.

It was completely electric.

In the nineteenth century, nobody knew the best way to design a car, let alone power one. Motorsport had only begun five years earlier, when the newspaper *Le Petit Journal* had organised the first race for 'horseless carriages' from Paris to Rouen. Lured by a purse of 10,000 francs, 102 eccentrics had entered cars straight from a steampunk fever dream. They were powered by everything from compressed air to propellers and hydraulics, with nine entrants claiming to be running on 'gravity'. Sadly, none of these wacky racers turned up on the day. Instead, the 26 cars that arrived were powered by steam or early petrol engines, with only 21 qualifying for the race itself. Even so, they made up a bizarre mix of concepts and ideas – from tractors to tricycles – and included seven-, eight- and even nine-seater wagons. Some of the vehicles had occupants seated up front facing each other, requiring the driver to peer between the passengers to see where they were going.

The eventual winner – despite taking a wrong turn and ending up in a potato field – was Comte Jules-Albert de Dion, who finished the 78-mile trip in 6 hours 48 minutes (not including a lunch break that lasted 1 hour 30 minutes). Sadly, the Comte was ineligible for the top prize as he'd hired a stoker to keep his steam boiler going. Instead, first place went to Albert Lemaître, who crossed the finishing line minutes later in a 3-horsepower, petrol-driven chassis. He had averaged a mere 11.5mph. Both men would soon find fame in other ways. In 1899, de Dion smacked Émile Loubet over the head with a walking stick. While aristocrats usually got away with that sort of thing, Loubet was the president of France, so de Dion ended up in jail for 15 days. Annoyed at how this incident was reported, once released, de Dion decided to publish his own newspaper, *L'Auto*, to set the record straight. It was a flop until 1903, when the staff held a bicycle race in an attempt to gain publicity and boost circulation. The race was the origin of the Tour de France. Lemaître, meanwhile, became infamous in 1906, when he overreacted to his wife Lucie's request for a divorce, shooting her dead before turning the gun on himself. He survived his own bullet to the head; at his trial the jury ruled it a crime of passion and he was acquitted.

Thus, motor racing began with disqualification for an aristocrat on a tractor and victory for a love-crazed murderer in a Peugeot.*

---

* This might sound bonkers, but it's standard fare for the history of Belle Époque France. The top prize for the Paris–Rouen was split between Peugeot (who came second and third) and Panhard (fourth and fifth). Panhard moved into military assault vehicles and no longer exists as a marque, while Peugeot is still going – unlike the car I bought from them.

Inter-city racing soon gave way to setting a land-speed record, a challenge that was met by the wonderfully named Marquis de Chasseloup-Laubat. The Marquis had helped establish the world's first auto club and was confident his electric Jeantaud Duc was faster than its petrol or steam rivals. While the Duc's chiselled nose meant it was fast, it wasn't particularly safe – it didn't even have brakes – and the Marquis wasn't willing to risk his own life to make history. Instead, he loaned the precious car to his brother Count Gaston de Chasseloup-Laubat who, on 18 December 1898, headed with witnesses to the open spaces around Achères, where Parisian sewage was dumped to fertilise beet fields. Here, they staked out a 1km course on the straight lane between the farms. It was a cold, wet winter's day and the timekeepers only had basic equipment that wasn't really up to the job. All we know is that Gaston covered the run in about 57 seconds. Rather than make a second attempt, the men, all soaked to the bone and frozen stiff, decided the time sounded good enough and made for the nearest bar. The land-speed gauntlet had been thrown down at a whopping 39.24mph.

The news made headlines around the world. Gaston was declared the 'Electric Count' and lauded as the fastest man alive – a title that conveniently ignored the fact trains had been going at double that lick for 50 years. The story had caught the eye of Jenatzy, an enterprising Belgian whose sharp features and brilliant ginger beard had earned him the nickname 'the Red Devil'. An engineer, Jenatzy had settled in Paris to manufacture electric taxi cabs and saw a world record attempt as an easy way to drum up business. He challenged Gaston to a race.

The duo met for their showdown on 17 January 1899. Jenatzy went first, setting a new top speed of 41.42mph.

Shortly after, Gaston made his own attempt, reclaiming the record by just over 2mph. Not ready to concede, Jenatzy arranged a rematch 10 days later, reaching a speed of almost 50mph; this time his rival failed to set a marker, with the Duc's motor burning out during its run. Now caught up in the competition, Gaston modified his brother's car and, that March, pushed the record to a staggering 57.65mph. The Electric Count was fastest once more.

But Jenatzy was tenacious. Realising he had to do something dramatic to beat his rival, he decided to create a car solely built for speed. The result was *La Jamais Contente*. The name had nothing to do with his pursuit of the record; it was actually a veiled insult directed at his mother-in-law.

Rolling up to the start on that April day, Jenatzy and his car made a unique spectacle for the crowd – mostly a gaggle of moustached men in top hats or straw boaters – who watched in awe. The smooth cigar-like body of the car – modelled after an airship – was built from a lightweight aluminium alloy called partinium, designed to protect the motor inside while shaving off weight from the frame. This sat atop a heavier cart chassis, while Jenatzy himself – dressed like a sea captain, with a flame-like goatee tapered to a fine point – perched high in a small driver's compartment at the back of his contraption.

With his head cocked imperiously to the timekeepers, Jenatzy took a deep breath, lowered his arm and gave the signal: 'Allez!' The stopwatch clicked; as it did so, Jenatzy flicked a switch and started his twin 25kW motors, each attached to one of the rear wheels, running at 200V and drawing about 68 horsepower to get the car moving. There was no roar of the engine or howl from an exhaust; the only clamour was a hummingbird whir from the engine and the crunch of Michelin tyres seeking grip on the road. The machine stuttered forwards slowly, gradually picking up momentum, before finally surging onwards.

Jenatzy later told physicians who asked him to describe the sensation of speed: 'The car in which you travel seems to leave the ground and hurl itself forward like a projectile ricocheting ... As for the driver, the muscles of his body and neck become rigid in resisting the pressure of the air; his gaze is steadfastly fixed about 200 yards ahead; his senses are on alert.'

At the controls, Jenatzy crouched down, trying to make himself as aerodynamic as possible. *La Jamais Contente* hurtled past the poop-fed farms on either side, wind whipping his coat and whistling through his orange chin decoration. Focusing his concentration, the Belgian kept the car almost perfectly straight, passing the men waiting for him at the 1km marker and pressing on. After another kilometre, he slowed the car and brought it to a gentle halt. The time was fast – but was it fast enough?

Driving back to the starting line, the Belgian soon had his answer. The crowd was cheering, feting him with wreaths and ribbons, and wrapping his machine in garlands. Jenatzy's wife climbed on to the back of the car with her parasol to pose for a photograph with her champion. *La Jamais Contente* hadn't just broken the land-speed record. It had reached a then-astonishing 65.79mph, making Jenatzy the first man to drive a car at a speed of more than 100kmh.

Camille Jenatzy had become the greatest racer alive. His feat was the perfect marketing tool – a demonstration of how electric cars were superior to combustion engines.

And yet it wasn't going to last. In under three years, *La Jamais Contente*'s record was beaten by a host of steam-driven and petrol-fed rivals. By 1909, the combustion engine had evolved to the point where steam couldn't compete either. Electric cars were gone.

By then, Jenatzy himself had abandoned his electric taxis to became one of the first motor racing drivers. He soon became the star of an emerging team from Germany, owned by the diplomat Emil Jellinek. Racers often used pseudonyms and Jellinek was no different, taking his daughter's first name in place of his own for the team. And so it was that, in 1903, Jenatzy headed to Ireland, where he won the Gordon Bennett Cup … for Mercedes.

Early motor racing was a risky profession and Jenatzy swore he'd die in one of the Mercedes cars. His prophecy came to pass – although not without fate showing its sense of humour. In 1913, while out hunting with friends, Jenatzy decided to prank them by crouching in a bush and making animal noises. Evidently his impressions were a little too on point, because one of his companions thought they'd found their quarry, spun around and opened fire. Jenatzy died on the way to hospital, carried there in the back of a Mercedes.*

The 1903 Gordon Bennett Cup is notable for another reason. As part of the competition, entrants from each country had to race under a chosen colour. The Americans went for red, the Germans white and the French blue. The British had originally raced in olive, but in honour of the host nation decided to paint their cars the deep, rich colour of Ireland's national symbol, the shamrock.

---

* There's some suggestion Jenatzy was murdered, given the man who shot him was the husband of Jenatzy's mistress. Either way, it's not smart to hide in a bush making animal noises during a hunt.

It was the birth of the tone that inspired this book's name: racing green.

★ ★ ★

I'm standing in the pantheon of the gods. In an ill-lit room, hushed to reverential silence, a semi-circle of eight Formula One cars are pointed at me, angled on a slight slope to show their entire frames. They span the ages, from the 1950s to the 2000s, their designs reflecting the cutting-edge science of each era. They're all a deep, instantly recognisable red, their maker's badge a modern icon. During the First World War, the leading Italian fighter ace was Count Francesco Baracca, who shot down 34 enemy planes before his own death in the mountains of northern Italy. After the war, his mother had attended a motor race and been so impressed by the winning driver she had suggested he adopt Baracca's personal emblem in honour of her son.

Baracca's symbol was a black prancing horse. The driver was Enzo Ferrari.*

The eight Ferraris in front of me are only responsible for a handful of the manufacturer's triumphs, a mere snapshot of its glory. This is the Hall of Victories in Maranello, Italy, the modern base of the world's most prestigious sports car marque and its legendary racing team. Behind me, an entire wall is occupied by trophies of gold, silver and sparkling crystal. Since becoming a founding team of Formula One in 1950, Scuderia

---

* Ferrari was driving for Alfa Romeo at the time; he only founded Scuderia Ferrari in 1929 and his cars didn't use the logo until 1932. When the team finally began to beat the Alfas, he described it as akin to 'killing my own mother'.

Ferrari has chalked up 16 constructors' championships, 15 drivers' championships, 237 race victories and 773 podium finishes. Ferraris have also won the 24 Hours of Le Mans outright nine times. It is a company whose sole existence is centred on racing, the desire to win etched into its DNA. The first Formula One car I saw live wore the Scuderia's sanguine sheen, as Felipe Massa hurtled through Silverstone's Abbey Corner in the wet and wild qualifying for the 2012 British Grand Prix. The cars here are still; my blood quickens.

It's the perfect place to start my pilgrimage into the science of motorsport. The future is, without question, green. The technology emerging from today's racers focuses on better, cleaner ways to power our world, and smarter ways to live our lives. Much of this journey will be about how motorsport found alternatives to the gas-guzzling machines that have helped push the world to the edge of climate catastrophe – and the astonishing tech already harnessed by teams like Ferrari that will help us adapt to what comes next. But first we need to go back in time.

Walking through Ferrari's museum halls, each one filled with gorgeous cars or dramatically backlit memorabilia, I'm reminded of a fist in a crimson glove. The cars are sleek, seductive works of aerodynamic art, but their engines are thirsty, merciless, savage beasts that belch throttled anger. These apex predators of the track didn't emerge overnight. They were born of a thousand design choices, introduced by engineers working on intuition, experience or the simple need to solve a problem. It's from this foundation – a constant need to think faster and smarter – that we came to a point where 20 cars going around in a circle can affect our lives so deeply.

The changes are writ large on the eight Ferraris lined up in front of me. Each one is more than a motor car; it is a hidden tapestry of paradigm-shifting thought that eventually trickled down into our own cars. They start with Mike Hawthorn's 1958 championship-winning 246F1, a long, aluminium chassis that looks like a kazoo on wheels. Hawthorn's triumph had come despite having complex hydraulic drum brakes, which could overheat or stop working in the wet. During the season, two of his teammates, Luigi Musso and Peter Collins, died after flipping their cars while taking a corner at speed.* The following season, the 246s used disc brakes – now standard on road cars, yet only adopted after Formula One showed their superiority. The new brake system saves lives to this day; scrapped drum brakes, meanwhile, have become little more than musical instruments for industrial bands. Who knew the kazoo played such a role in heavy metal?

A few cars along, and Niki Lauda's alien-looking, low-body 312T shows how aerodynamics had come to the fore, even if they were not fully mastered. The most successful Formula One racer in history, the 312T has a turbocharger, five-speed gear box, wing mirrors and triangular NACA ducts – little pressed-in pyramids to channel air to the engine – stolen from fighter jets. It isn't all modern, though; part of Lauda's success was down to the car's independent suspension – a design based on the very one used by Comte de Dion's steam tractor. Sometimes going forwards means looking back.

---

* The 1958 season was the bloodiest in F1 history, also taking the lives of Stuart Lewis-Evans and Pat O'Connor. Hawthorn retired after his victory, but died in a road traffic accident three months later, aged 29.

Finally, past the similarly dominant chassis that brought Jody Scheckter victory in 1979, sit the cars of the legendary Michael Schumacher and the acerbic 'Iceman' Kimi Räikkönen, the last Ferrari drivers to clinch the title. The metal chassis has evolved to use lightweight carbon fibre as well. The open cockpit has become more refined, the car shape more familiar as designs were honed and perfected. The stick shift and clutch pedals are gone too, replaced by the 'flappy paddle' gearbox first trialled by Ferrari in 1989.

But these tweaks aren't the reason I've made my journey. I'm here for what lies under the red paint: the engines. Ferrari is the undisputed champion of twentieth-century racing, but the precision-tooled machines that powered its victories belong in the past. Combustion engines may have ruled our roads for more than 100 years, but their days are over.

I've come to Maranello to say goodbye.

★ ★ ★

An internal combustion engine, or ICE, runs on exploding dead sea critters. Petroleum, crude oil, is the liquified remains of zooplankton and algae, buried under sedimentary rock and subjected to heat and pressure over millions of years. Now, all that remains of these prehistoric beasties are hydrocarbons – linked carbon atoms with hydrogen molecules on the side. Different arrangements of chains, and different lengths, cause your fuel to burn faster or slower. But it's still just dead prehistoric stuff that's been refined into particular mixes.*

* Crude oil isn't all one substance, but bears the unique fingerprint of how and where it was created. It can include bits of other stuff too, such as a small pinch

If this description of petrol sounds unscientific, keep in mind that fire is complicated; ask any combustion expert and they'll tell you we still have *absolutely no idea* what goes on at the heart of a candle flame. If you take the simplest hydrocarbon (methane, CH_4) and react it with oxygen, you still have several hundred possible products and by-products, all being created through hundreds of different chemical reactions, running at different rates that can vary in a moment as the environment changes. And engines run on much larger hydrocarbons than methane – octane, one of the main components of petrol, is C_8H_{18}. In fact, ICE fuels contain a host of different hydrocarbons, configured in many different shapes. These all react while being mixed with fluctuating amounts of air and waste gases in a turbulent, ever-changing maelstrom of temperature and pressure. In motorsport, nothing is simple and everything needs a caveat. So, if you want to talk generally, 'exploding dead sea critters' is the best you can do.

The modern ICE came about in 1876, when former travelling grocery salesman Nicolaus Otto came up with the four-stroke cycle. In each engine block are a series of holes, called cylinders. Inside these cylinders, a piston is connected to a rotating crank. First, the piston moves down. This sucks in a mixture of fuel and air. Next, the piston moves up, compressing the petrol and air at the top of the cylinder, where a device called a spark plug causes – unsurprisingly – a spark. In an instant, the petrol in the cylinder ignites, ripping apart those long hydrocarbon chains, releasing energy in the form of a miniature explosion and creating waste gases. The

of dinosaur for flavouring, although the chance of a *T. Rex* in your car is slim to none.

gases expand, pushing the piston back down the cylinder, creating the power that turns the crank and drives the machine. As the crank turns around, it pushes the piston back up, forcing the gases out of the engine as exhaust and starting the whole cycle again.*

The shape of the engine block matters. Formula One cars use a V-shaped engine, which puts pistons on either side, grouped in pairs and tilted towards each other at a sharp angle (F1 engines have their pistons at 90 degrees). The result is power strokes coming from two different directions, each hammering home with a force equivalent to four elephants pushing down. A Formula One car can operate at up to 15,000 revolutions (turns of the crank) per minute and has six cylinders; its engine contains about 50,000 explosions over the course of a single lap and 30 million over its lifetime. (In the late 1990s, when there were no restrictions on rpm and engines could have up to 12 cylinders, this number was far higher.) This means the pistons have to move rapidly, with an acceleration equivalent to 9,000 times gravity, requiring specially designed engine lubricants to protect parts and prevent energy lost through friction. The peak temperature inside one of these beasts reaches about 2,750°C and they're so finely tuned that they can't be turned over cold: they need to be warmed up with oil and water first. A Formula One engine is the art of raw power.

Sadly, as you might expect from anything that involves blowing up prehistoric goo, an ICE is really inefficient. Although hydrocarbons contain a lot of energy, in a standard

* Diesel engines are completely different, but that doesn't really matter – we're ditching fossil fuels after this chapter.

car only around 35 per cent of it goes into pushing the piston down; the rest is lost. This is called an engine's thermal efficiency; the more efficient it is, the further you can go and the less waste is produced.

Energy is never created nor destroyed, it's only ever transferred into another form. In an ICE, the waste is converted to heat, half of it passing into the engine block (making things hot) and half of it carted away in the waste gases caused by incomplete combustion. Sadly for the planet, these gases include such delights as more hydrocarbons and particulates, carbon dioxide (CO_2), carbon monoxide and a whole assortment of nitrogen oxides, which are so numerous they're usually just referred to as NO_x. None of this is particularly nice stuff and represents a major part of our global carbon footprint – a measure of greenhouse gases in our atmosphere that cause climate change. According to the International Energy Agency, transport accounts for around a quarter of global CO_2 emissions, with a whopping three-quarters of it (some 6 billion tonnes of CO_2) coming from road traffic. Most of that is from the ICE.

It would be wrong to think that ICEs haven't evolved, though – and motorsport has been leading the way throughout. In 1912, Peugeot's engineers came up with an idea so radical they were dubbed *Les Charlatans*. Rather than have two valves, one to let air into the cylinder and one to let air out, the canny Peugeot team used four valves (two for air in, two for exhaust out). This maximised surface area, meaning more air could come in to react with the fuel, and more exhaust could be ejected, increasing power and reducing waste energy. When Peugeot took its car to the 1913 Indianapolis 500, driver Jules Goux was so confident of victory he and his mechanic began drinking champagne *during* the race, swigging it as they hurtled around the track. The duo managed to polish off four bottles between them before the chequered flag, finishing 13 minutes ahead of their nearest rival.*

ICEs are still being honed and improved, with engineers, physicists and chemists working together to produce some pretty smart innovations. In 2017, the Mercedes-AMG Formula One team announced something previously thought impossible – an ICE that had more than 50 per cent thermal efficiency. Normally, although not always, an engine wants to have exactly the right amount of fuel and air entering the cylinder to react with each other. A rich fuel mix hasn't got enough air and, although it produces more power, it is inefficient. Conversely, a lean mix has too much air; this is a more efficient way to burn, but is hotter and produces less power. Rather than try to balance their air and fuel, the Mercedes team did something different: in each cylinder, they built a small pre-chamber containing a rich mix, while leaving the main chamber with a lean mix.

This was a brilliant innovation. Inside the pre-chamber at the top of the cylinder, the rich mix was ignited by the spark plug. This produced lots of power and caused jets of flame to shoot out of specially designed holes. The flames then ignited the leaner mix, forcing the piston down in a smooth, rapid burn. By deliberately unbalancing their

---

* Goux's pre-win drinking was only the second weirdest thing to happen that day. In the race's closing stages, the car of Charlie Merz caught fire. Rather than give up, Merz decided to keep going. He finished the race driving a fireball, with his mechanic Harry Martin sprawled out on the bonnet trying to beat out the flames. Despite their near-death experience, Merz and Martin finished in third place.

chemical equation, the Mercedes team were able to get the best of both worlds, making their engine powerful and yet more efficient overall.

But fuel mixes and chambers were only part of the Mercedes process. Modern F1 engines also contain a turbocharger. This is basically a turbine that is made to spin as exhaust whips out of the engine, powering a compressor that sucks in extra air for the engine's cylinders. Mercedes had connected this turbocharger to an electric generator called an MGU-H (motor generator unit – heat). Along with energy recovered from brakes (which we'll come to later), excess energy was then recovered by the car and stored to be used later. To make things even fancier, Mercedes had split their turbocharger in half, which kept the heat-sensitive compressor cooler and made it more effective.

Although this is incredible hybrid engineering, it means an ICE still can't pass 50 per cent thermal efficiency on its own: you need some help from a turbocharger and an electric battery. Mercedes' feat is just the ICE's swansong.

In Maranello, stepping out into the morning chill of an Italian winter, an uneasy hush lingers around Ferrari's factory. There's no throaty roar from a V12 engine; there's not even a quiet whisper from the ghost of the great Enzo to bid me farewell. Ferrari is the last vestige of a dying breed of manufacturers. Within a decade, ICEs will begin to vanish from our roads. Countries including China, India, Japan and the UK have already put together plans to phase out combustion engines in the next 10 years, and manufacturers such as Volkswagen and General Motors plan to discontinue their lines. Formula One itself has announced it wants to be completely carbon-neutral by 2030. It can't do that with conventional fuels in an ICE.

For all that 50 per cent thermal efficiency is a technical milestone worth celebrating, the cars of the future need a viable alternative to fossil fuels.

The good news is they have one. It's just taken us a century to come full circle.

CHAPTER TWO

Speeding Bullets

The Bonneville Salt Flats stretch out like a crack in creation. Standing on the flats (also known as the salt pan), the world reaches out as emptiness until its hazy vanishing point, where the soft crunch of white beneath your feet surrenders to distant tapered peaks and the open, empty blue of the Utah skies. It is a realm of nothing, as if you are in a computer simulation waiting for the universe to load. It's where heroes come to roll dice with the gods of speed.

In 1914, the now ghost town of Salduro lured daredevils to the flats with a promise: '[The] fastest machines in the world will compete for the world's record on the famous salt beds, which afford the finest races in America. No dust.' And so it proved. As the land-speed records began to climb, poo farms outside of Paris were no longer suitable. You needed somewhere long, straight and flat. Beaches became popular, such as Daytona in Florida in the US, or Blackpool and Pendine in the UK. In 1935 it was Bonneville's turn, with Sir Malcolm Campbell's *Blue Bird* becoming the first car to surpass 300mph. In 1947, Bonneville became home to the first single run beyond 400mph, set by John Cobb and the *Railton Mobil Special*; and in 1963, Craig Breedlove and the *Spirit of America* brought the record into the realm of jet turbines.

Each year, the Bonneville Speed Week acts as a festival for these acolytes of Mercury, their triumphs heralded by the colours they wear. 'If you go less than 100mph,' one guru of the pans told me, 'just get off the salt flats. At 200mph, you're official: they give you a red hat and everyone knows you're a land-speed racer. At 300mph the colour of the hat changes to blue. Now, you're in a pretty small club – maybe 80 people. And when you go over 400mph, the hat becomes black. You're in a club of seven. And you're fast by any standard.'

Early morning on Tuesday 24 August 2010, David Cooke was staring out at the Utah horizon, dreaming of glory. Or, maybe, just his bed. He was fatigued, deprived of sleep beyond reason. The team had worked for a year on the details of his planned race attempt; now they had to deliver. That meant 24-hour days at the mercy of the salt flats. For all its ethereal beauty, Bonneville is unforgiving. Thunder clasps the skies and rain falls, turning the blanched playground of racers into a reflection pool, dampening the salt to mush. Some years, the rains cause the nearby hills to liquify and smear the track with mud. At other times, the wind howls across the plain, stinging your lips, caking your engines and shattering your dreams. Even if the weather holds, you only have a small window in which to make your run – and if something goes wrong, there's no garage to get another part. 'You work all year, you kill yourself working late nights, you spend hundreds of hours on tiny little problems,' Cooke remembers. 'And if one of them fails, the deal's off.' Those hundreds of hours weren't from career professionals, either; they were spent by students, finding time between study sessions to work on their passion.

At 24 years old, Cooke was one of the youngest team leaders in the desert and the crew who elected him – comprised entirely of classmates from The Ohio State University* – the

* I know this looks like a typo, but the university is *The* Ohio State and very particular about it – so much so in 2019 it tried to trademark the word 'The'. It was not successful.

youngest group assembled. They were also one of the most storied. In 2004, his car's predecessor, the first Buckeye Bullet,* had set an international speed record for an electric car at 271.74mph. At the 2009 meet, Cooke's first season in charge, the Buckeye Bullet 2 had pierced the 300mph barrier, using hydrogen fuel cells.

This time it was the Buckeye Bullet 2.5. Same chassis; same aerodynamics, designed by Kim Stevens, a promising engineering student. But now, with the help of electric car manufacturer Venturi, it had been converted to electric: *La Jamais Contente* reborn. If the run worked, it would be the first electric car in history to become one of Bonneville's blue cap legends – and push the world record for electric racers into a realm no one thought possible.

But the salt pan doesn't care about your backstory. A year earlier, under the watching eyes of a documentary crew from the Discovery Channel, Cooke and the team had faced countless disasters. First an inverter had failed, prompting someone to fly out with a replacement. A panel came loose, forcing the team to fasten down their aerodynamic chassis with roll after roll of tape. Then a problem with their motor had ripped off its internal insulation. 'The local company told us they needed two weeks to fix it,' Cooke recalls. 'I said "Can you do it in 30 hours?" They asked my budget. I said "Let's pretend I don't have one."' To this day, he still doesn't know how the team pulled it off, beyond 'a lot of people working their butts off'. That day, even the track marshals were on side;

* The first *automotive* Buckeye Bullet; the original bearer of the moniker was Jesse Owens, the black athlete who starred at the 1936 Olympics and embarrassed Nazi Germany.

when the team fell short of 300mph by a whisker, the first thing they heard over the radio from control was 'Everybody stay in place, we're getting this damn record!'

Now converted to the Bullet 2.5, Cooke was hoping they wouldn't face a similar problem. The Bullet was required to make two speed runs along a 12-mile course, once in each direction within 60 minutes, in order to be considered for a world record. The timekeepers would take the two speeds, average them out and compare them with previous holders. At the starting line, Cooke and a handful of others leant their weight against their car, pushing it out to the track; the rest of the team were miles in the distance at the finishing line, ready to cool the car down and prepare for the run back. Inside, veteran driver Roger Schroer, the man who had piloted their previous attempts, made his final preparations.

Then they went. Not fast at first. Steady, roughly the same acceleration as a Honda Civic, building up speed gradually, agonisingly. Yet while a Civic would have begun to top out and slow, the Bullet kept going. The same acceleration, up and up and up … until it was a speck on the edge of reality.

'The heaviest moments in your life on your heart are after the car leaves,' Cooke recalls. 'You can't see it any more, it just goes over the horizon. And then you just have the radio, and they don't say a lot. You just get a safety instructor: "Through mile one. Through mile two … " You're just praying you don't hear "fire" or "tumble". Then the other half of your heart is waiting to hear the speed.'

The Buckeye Bullet 2.5 never drove again after that day. The clutch went, the torque (turning force) from the motor ripping apart the half-inch steel teeth that connected to the gearbox. After yet another sleepless night of repairs, the team admitted defeat.

It didn't matter. On their two runs, they had averaged 307.7mph. The Buckeye Bullet 2.5 was the fastest electric car on Earth.*

★ ★ ★

If the world ends thanks to climate change, you can blame Ohio. The Buckeye State has a long and sketchy history of trashing the planet. As the centrepiece of the US Rust Belt, a nexus where heavy industry and coal mining meet, Ohio's manufacturers happily churned out some of the most uneconomical, downright disastrous creations the world has ever seen while turning its cities into a fetid, soot-choked sump. Between 1868 and 1969, the Cuyahoga River in Cleveland was so polluted it *caught fire* at least a dozen times, a situation so bad it prompted Richard Nixon to found the Environmental Protection Agency. In 2014, half a million people around Toledo were told not to use, never mind drink, their tap water: decades of chemical run-off had turned Lake Erie green with algae. But none of that even comes close to the damage wrought to the planet by one man: Thomas Midgley Jr.

Midgley grew up in the state capital, Columbus. His father was an inventor and evidently the need to create rubbed off on his nascent mind, because in 1911 he had graduated from

* The fastest electric car *anywhere* is a Tesla Roadster Elon Musk shot into space in 2018. At the time of writing, it was orbiting the Sun at about 6,483mph – almost 8.5 times the land-speed record.

Cornell with a degree in mechanical engineering. Five years later he began working for General Motors in Dayton and was tasked with solving a problem called 'knocking' in the company's Cadillacs. This is basically an engine's version of a heart murmur, and occurs when mini pockets of air and fuel explode on their own, separate from the cylinder's four-stroke cycle. Instead of a perfectly timed ignition from the spark plug causing a lovely, smooth flame that pushes the piston down, the fuel burns unevenly, causing shockwaves that rattle the engine. Knocking isn't really a problem any more, although it's why there are regular and super unleaded pumps at the petrol station – the higher the fuel's 'octane rating', the more compression a fuel can take before detonating and the lower the chance of knocking.* Combustion engines in the early twentieth century weren't reliable at the best of times; knocking made them shake apart.

Midgley decided to solve the problem with chemistry. He started experimenting with a host of different substances in his fuel mix, trying to find a combination that prevented the dreaded knocking. A lot of this was pure trial and error – at one point he added butter into petrol to see what would happen – but eventually he hit on a solution. If you added tetraethyl lead (TEL) into petrol during combustion, this would be broken up to form lumps of lead and lead oxide gas. This didn't just eliminate knocking, it improved engine performance, too. Midgley had just discovered leaded petrol.

---

* There's no such thing as a standard octane rating, as it varies by country and measurement. Mountainous states in the US also have lower standard octane ratings than those at sea level as their air is thinner and knocking is less likely. Diesel cars want a *low* octane rating as their engines operate differently.

Lead is really bad for you. It causes everything from headaches and seizures to mood swings, developmental delay and premature birth; some have even linked it to increased aggressiveness and violence. Although Midgley insisted it was safe, General Motors cynically decided to drop all mention of the metal and market the new product as Ethyl, hoping that nobody would notice. Unfortunately, the name change didn't prevent lead's effects. At the Standard refinery in New Jersey, 17 workers died after going violently insane when making TEL. Nearby, workers at DuPont called their site the 'House of Butterflies' because they kept hallucinating they could see non-existent insects. The *New York Times* argued the deaths were 'not sufficient reason' to stop making leaded petrol, given it was such a money-spinner. It wasn't until the 1970s that science accepted it was a problem; by then, Midgley's legacy was to turn every car exhaust into a lead spray can, coating streets with a fine dusting of heavy metal for the locals to breathe in.

Today, leaded patrol is banned in every country around the world and has been largely gone from US streets since the 1980s (except in NASCAR, where leaded petrol was still used until 2008 – but we'll come to that later). Even so, there are an estimated 15 million people in the world with brain damage from lead exposure, the vast majority of whom can thank petrol fumes. In 2011, the United Nations estimated that banning lead petrol saved 1.2 million lives and around $2.4 trillion in costs *every year*, as well as linking the ban to raised intelligence and fewer crimes.

As if giving the world a hefty dose of lead poisoning wasn't bad enough, Midgley hadn't finished having bright ideas. After a short holiday (to recover from lead poisoning – perhaps linked to him huffing TEL in front of journalists to prove it was safe), Midgley moved into refrigeration and created a new chemical called Freon. It was the world's first chlorofluorocarbon, or CFC. While these are fantastic for freezers and aerosol cans, in the 1970s scientists realised the lighter-than-air gases were drifting up to the Earth's stratosphere, where they were being broken down by ultraviolet light. This set loose chlorine atoms, which bonded with ozone and burnt a hole in the Earth's natural atmospheric sunscreen. In 10 years, Midgley had laid the groundwork for leaded petrol and the hole in the ozone layer: the two greatest ecological disasters of the twentieth century. The man was an environmental wrecking ball.*

It's only fair, then, that Ohio – and Columbus in particular – does its share of the global clean-up. It's why I'm standing on a quiet side street, not far from the Olentangy River and the massive 100,000-seater Ohio Stadium. The Ohio State University is a city to itself: 68,000 students and 35,000 staff populating world-class hospitals, supercomputer centres and college programmes. It's so large it even has its own airport, with 70,000 flights a year. Behind me, on the fringes of the campus, a typical suburban lane is filled with picturesque wood-panelled houses in pastel shades, complemented by white picket fences and leafy, well-tended lawns. Across the road is more industrial – a nondescript red-brick building. It's The Ohio State University's Center for Automotive Research

* Midgley came up with a third invention of note. In later life he contracted polio and invented a system of pulleys to haul himself out of bed. In 1944, he was using the machine when he slipped and strangled himself to death. It was a fitting end to a man whose genius lay in coming up with the worst brilliant idea possible.

(or CAR, see what they did there?). Despite appearances, it's one of the most advanced motoring facilities in the world.

The CAR aims to be a one-stop shop for the automotive industry. Here, or on the nearby test track, you'll find research into everything from autonomous vehicles to fuel consumption, advanced modelling and even underwater robotics. Each room has something even more incredible. When someone needs to study combustion in an engine, CAR is the place to do it; they have a camera capable of shooting 30,000 frames a second to film the flame creation and burn, allowing them to see exactly what happens in an engine's cylinder. They have just about every test you can run on a motor, even down to a car's cybersecurity. They deal with the smallest chassis to semi-trucks and highway haulers, and just launched a programme to improve the US government's transit buses. It's the kind of facility that makes a petrolhead drool.

Pass through the building, though, and you'll come across something more remarkable still. Sat tight against the wall is a long, thin box almost 10m in length. Painted scarlet and grey, its nose is hooked down to the ground, with a small bubble of a cockpit at the rear. It's empty now, like a snake that's shed its skin, leaving only a hollow carbon fibre shell. It's the remains of the first Buckeye Bullet, 15 years out of time. A strange fate for what was, in its heyday, the fastest electric car in the world. But the story of CAR's electric racing doesn't begin – or end – there.

★ ★ ★

For 100 years, electric cars were a punchline, the fodder of tongue-in-cheek news reports that suggested we'd all be driving around in ridiculous bubble cars. That doesn't mean there weren't attempts to take them seriously. In 1968, MIT and Caltech held the Great Electric Car Race between each other's campuses, each on opposite coasts of the continental US. MIT arrived first, but was penalised for towing its car for 37 hours, giving the win to Caltech. But the race was the exception rather than the rule.

Frankly, it's because batteries were rubbish.

We don't really know when the battery was created. The oldest thing to claim any similarity, the Baghdad Battery, comes from Iraq and dates from anywhere between 250BC and AD650. It's hard to date because it's little more than a ceramic pot with a copper tube and an iron rod in it, with acidic marks on the metals suggesting wine and vinegar were in the jar too. Whether it's a battery or not is unknown. Perhaps it was used for electroplating, or giving someone a zap; perhaps it's just an odd coincidence. True batteries didn't emerge until about 1800, when Italian physicist Alessandro Volta wrote to the Royal Society in London describing the result of a series of experiments he'd been conducting (literally!). His 'voltaic pile' was little more than a stack of discs made from copper and zinc, spaced out with brine-soaked cloth. But Volta's metal strips and salty rags would change science forever.

Batteries are relatively simple. They are made up of cells, which have three basic parts: two plates, the anode and cathode, and an electrolyte in the middle. When you create a circuit, a series of chemical reactions occur (exactly what depends on what you're using to make the battery), causing negatively charged electrons at the anode and positively charged ions at the cathode. The electrons want to get to the ions, but are blocked by the electrolyte. That means the only way around is to go via the circuit – a kind of traffic detour – generating a current and powering your device.

Once the chemical reaction is complete, the cell is useless: there's no more traffic that needs to flow along your circuit. This is where rechargeable batteries come in; here, when electrical energy (*i.e.* from a socket) is applied to a cell, the chemical reaction is reversed and the cell is recharged. And in case you're wondering why we call them 'batteries', the name comes from none other than US founding father Benjamin Franklin. In 1748, Franklin was playing around with Leyden jars – glass containers capable of storing an existing charge – and hooked up 35 of them together. As they all discharged at once, it reminded him of cannons firing in unison. Eventually, 'battery' came to mean anything storing electricity in a chemical form.*

When it comes to transferring energy, batteries are really efficient as there's virtually no waste: the chemical energy becomes electricity. While an ICE only delivers about 35 per cent of its energy to power a vehicle, a battery delivers about 90 per cent. Better still, there aren't any waste products such as exhaust fumes. The downside is that the amount of chemical energy stored in a battery is nowhere near the same level as that trapped within a fossil fuel. The energy by mass of petrol is more than 12,000 Watt-hours per kilogram (Wh/kg). For a lead–acid battery, such as the one Camille Jenatzy used, it's about 30Wh/kg.

---

* The artillery battery originates from the old French *battre*, meaning 'to strike'. Oddly enough, today we wouldn't call Franklin's effort a battery at all; as the energy was stored as an electric field, it would be classed as a capacitor.

This is where The Ohio State team entered the story, explains CAR director Giorgio Rizzoni. Cooke and I are now in his expansive office, trophies on the wall, a photo of the team with President Obama taking pride of place. Lunch is a hefty plate of sandwiches, followed by coffee. 'I'm Italian,' Rizzoni smiles. 'It'd be un-Italian *not* to have an espresso after lunch.' With a constant, cheeky grin on his face, Rizzoni literally wrote the book on electrical engineering. Throughout the 1990s, he says, electric cars had seen a resurgence. In the US, the state of California enacted a Clean Air Act demanding car manufacturers produce zero-emissions vehicles. The big problem was that no one was thinking electric first; rather, they were retrofitting cars in a way that just didn't work.

To solve the fundamental problems posed by electric engines, teams started racing cars to spot the areas that needed development. 'In the summer of 1994, we had our first race,' Rizzoni says, referring to The Ohio State's student electric car team. 'But there were lots of mistakes made by people designing the chassis. For example, the brakes were completely wrong; people who designed them made them for a 1,500kg car, much heavier than a race car. The driver couldn't put enough force on the pedal to stop. So, we were involved in a lot of arguments about how to improve.'

The arguments resulted in The Ohio State winning three national championships. But design flaws aside, the big problem was always the sheer weight of the batteries. 'When we had to do a pit stop, we had to replace 600kg of lead–acid batteries in 30 seconds,' Rizzoni recalls. 'Typically, we had four students each side, each pulling out two boxes [weighing 30–40kg]. The pit crew had to be a robust group of young men.'

The lessons learnt from racing led to the next step: Bonneville. Inside the first Buckeye Bullet were 15,000 nickel-hydride C batteries – the kind you might use for a torch or toy. 'It was the equivalent of 16 Toyota Prius battery packs, completely reconfigured,' Rizzoni recalls. 'Fabricating this was ridiculous. Just think about all the metal around the batteries – just the skin that's not really doing any work. There had to be a welded spot at each of them.'

The next generation, Buckeye 2.5, was just as challenging. By then, battery technology had evolved and the team had started using lithium–ion batteries. But that didn't reduce the complexity. 'Everything in the powertrain was a prototype,' Cooke chips in. Since his adventures in the desert, he's graduated to senior associate director of CAR. 'The batteries are based on production technology, but they were a one-off. The inverters were the dream of a new product that we validated. The motors were production, but implemented in a totally different way. On board, we have about three miles of control wire, thousands of signals and huge amounts of power.' The price tag for the car was around $5 million.

This design created unique problems. 'You started getting electric noise,' Cooke explains. 'We were at the point where things would shut down because of the interference. We spoke to an expert who said you need to space things apart by three feet [to prevent noise] – but we had to use a tunnel between the wheels that was eight inches. It's these little things that hold you up – it's not the motor, it's things nobody ever dreamed of.'

Yet these are the types of puzzle from which engineers are born. 'When you're 18 and you are faced with a crazy challenge like this you get pushed to a whole new level,' Cooke says, Rizzoni nodding in agreement. 'I come from a small rural community. I was a good student at high school but engineering was a stretch for me, maths was real tough. When I went to class at university, focusing on trying to understand what they were throwing at me, it didn't make sense. I was killing myself with the curriculum. But I came here, and you're faced with real problems. You solve a problem and go, 'Well, I learned about friction today', or fluid flow, or thermal cycles. It becomes a magical connection. I wouldn't have become an engineer without this.'

Cooke is far from the only graduate in this school of excellence. Kim Stevens – one of the original Buckeye Bullet team who went on to design Buckeye Bullet 2 – is now trackside aerodynamicist for the Mercedes team that has dominated Formula One for almost a decade; one of the earlier racers, Jackie Marshall, became chief engineer for Ford's F-150 and Expedition SUVs. Of Cooke's 12-strong Bullet team from 2010, members have gone on to NASA's Jet Propulsion Laboratory to work on the Mars rovers, design planes for Boeing or the military, and one even works at Apple. 'The next generation of the iPad is something a guy with an automotive background has worked on,' Cooke grins. 'They're all at interesting places in life.'

Beyond training, centres like CAR also start the ball rolling for technologies that automotive giants consider too risky. They are the step between blue-sky thinking and on-track technology, which in turn leads to what we drive on our streets. 'What you have to realise is that from laboratory demonstration to something robust enough to put into production is a long road,' Rizzoni says. 'The batteries that are in today's production models were made three to five years earlier. And we're still not finished. Right now, we're trying to figure out the next record. We're looking at different batteries, trying to remove a few hundred kilograms, finding the right technologies, trying to get 2MW of power – right now we've reached about 1.6MW. If we can do all that … ' Rizzoni pauses. 'In history, there have only been seven wheeled vehicles that have gone over 400mph. If we could become the eighth and the first to do it with an electric vehicle? I think that would be pretty damn cool.'

Jamais Contente. Never satisfied. And the latest iteration of the Buckeye Bullet is only 50m away.

★ ★ ★

Lunch finishes, as Rizzoni promised, with a civilised espresso. An hour later, we're standing in the workspace – effectively a giant shed – for The Ohio State University's student teams. There are eco cars, motorcycles and racers, each in their own bays. Toolboxes, cabinets and shelves stacked with metal parts are lined up among grease-covered pits, stacks of tyres and even speakers for a little entertainment. In the rafters, hanging among scarlet-painted girders, are team pennants and flags for a little pep spirit. But the object that draws the eye is parked in the centre of the warehouse.

Buckeye Bullet 3. With the help of Venturi and still driven by Schroer, in 2016 the team set the current electric land-speed record: 341.4mph. Four-wheel drive, its two motors powered by 2,000 cells and generating 3,000 horsepower, this is the most potent electric car on Earth. Today, its aerodynamic skin is off, revealing the honeycomb structure that keeps its

3.5 tonnes from spilling across the salt pan. One metre wide and more than 11m long, it looks like some kind of futuristic behemoth. It reminds me of an unmasked Terminator.

'The frame is chromoly steel,' Cooke says, referring to the shell. An alloy of iron with chromium, molybdenum, carbon and manganese, chromoly is far stronger and harder than normal steel: it's used in everything from roll cages and commercial aircraft to the barrel of an M16 rifle. 'The tyres are specialised for land-speed racing – there's practically no rubber on the outside of it, a 32nd of an inch. In racing, that would be shredded off.' The cockpit is notable too, designed to enable a swift exit should an accident occur. At Bonneville, Schroer has to demonstrate that, in an emergency, he can hit the fire suppression system and parachute, bring the car from nearly 400mph to a standstill and then scramble free from the cockpit in under 14 seconds.

I walk over and peer inside the monocoque, sandwiched up near the tail fin. It looks tight. 'You want to climb in?' Cooke says.

'Uh, hell yes!' I take a deep breath. At 6ft 5in and packing a hefty beer gut, I'm not designed for such spaces. I follow my guide's instructions. One foot over the frame … and another … turn to face forward … shimmy over the GPS … twist your knees out and feet in and slide down until … 'Uh … I'm stuck.'

I pause, trying to wiggle out. My legs just won't bend under the lip. Slowly, a crowd of students begin to assemble, abandoning their projects to watch my attempt to pull free.

'I don't want to make you feel bad,' calls out one, 'but a 70-year-old man does this in under 10 seconds.'

'At least you didn't rip your pants,' another adds.

I half-smile, wiggling my butt for freedom. Somewhere, a belt buckle kisses a cheek. With a grunt, I grip the steel frame and yank my legs free, near-falling into the cockpit's seat. I'm in.

I take a moment to compose myself. Then, with awkward embarrassment, I lift myself out. 'Thanks for that,' I say, blushing. There are grins all round.

'Don't feel bad,' Cooke says. 'I can't do it either.' It turns out *most people* can't. I shake my head and decide that I'm not built for setting a land-speed record. Gradually, the crowd disperses and I'm left with Cooke by the Bullet's side. The car won't be at The Ohio State for much longer; it's soon destined to head over to Venturi's headquarters in Europe, where it'll go on display along with the company's other triumphant projects – electric bikes and concept cars. 'Hey,' Cooke says, 'you should go over to Monaco, see what they're doing there.'

I'd love to – and fortunately I'm already on my way to Europe. My next stop after North America isn't Monaco, though. It's Spain – and the motor racing series that forms the next step in a chain of innovation that will bring commercial electric vehicles closer to the mainstream.

It's time to enter the world of Formula E.

CHAPTER THREE

Together in Electric Dreams

'Sport and war,' a voice murmurs beside me. I'm standing on the pit wall, watching the daredevils that fulfil Jenatzy's legacy swoop past at 174mph. 'That's where most innovation comes from, right? Sport and war. I always liken what we do to the aircraft industry in a time of war. The rate of development is massive.' I turn to the man beside me – Chris Gorne, the technical director of the Envision Virgin Formula E team – and smile.

There are numerous strange links between both. One of the most popular automotive engine lubricants is the German Liqui Moly; it was invented by First World War fighter pilots who noticed that adding molybdenum disulphide (MoS_2) to their motor oil meant their engine didn't stop if it suddenly lost pressure. This allowed them to land safely if their tank was hit. It works the other way, too. In 1913, Mercedes' grand prix car used an aircraft engine to outpace its rivals. One of the cars was in a London showroom when war broke out the following year, so the British confiscated the engine and sent it to Rolls-Royce. The result – the Rolls-Royce Eagle – powered early fighters and bombers. After the war, the first non-stop flight across the Atlantic – in 1919 by John Alcock and Arthur Brown – would be completed on the back of two Eagle engines.

I'm about to reply when my words are whisked away. As if on cue, yet another adventurous Belgian in a Mercedes – this time Stoffel Vandoorne – howls through the pit lane, ready to test and tweak his team's designs. I'm still jetlagged after my North American adventure, but it snaps me back to the present in an instant.

A Formula E car sounds like a missile.* It approaches as a soaring, roaring whoop, whips past like an electric-powered sucker punch, vanishes as sonic neat alcohol evaporating from your tongue. The closest experience (unless you want to be on the receiving end of the latest military hardware) is to stand on a platform as a high-speed train blasts through the station. Strip away the train's inelegant clanks and give it the ability to decelerate to a standstill in moments … and it would still be a pale facsimile. A Formula E car is only 10 decibels louder than a family saloon – no need for the ear protection demanded by Formula One's hungry, hungry hybrids – but anyone who thinks motorsport needs to deafen you to leave an impression can think again.

The Circuit Ricardo Tormo is about 12 miles outside Valencia, nestled among the characteristic canvas of the Iberian Peninsula. It is a world of gentle hills, scrub bushes and sand under soft sun and open skies. The track itself is in the shape of a sandal shoe: a long, 876-metre straight that rushes out to the far extreme, turns up to create the back of the shoe, twists back towards the straight, then forms a hook that weaves in on itself before finally pushing out into an arc that bends and completes the loop. On its north side, curving with the bend of the track, is a 60,000-seat grandstand; to the south, the pits and the main building. It's named after the local sports hero, a two-time MotoGP world champion who died from leukaemia

* You might be wondering how I know what a missile sounds like. Let's just say I've had an interesting life.

shortly before it opened in 1999. It remains a key fixture in the motorcycle racing world, although it's probably most famous for the activity that goes on between race days. Its European location and clear weather make it the ideal proving ground for the fastest sports in the world.

Today, the Formula E circus is in town for pre-season testing. At the back of the paddock, a maze of stacked parts, cables and towering lorry trailers are set up to power garages taken over by the teams. On the grandstand, journalists line up to watch the telemetry reports, checking out the runners and riders of the season to come. When Formula E began in 2014, it was laughed at as a side attraction, something that would never supplant the crown jewels of motor racing such as Formula One or Le Mans; a sport derided as *Super Mario Kart* by its critics for gimmicks such as 'attack mode' and 'fan boosts'. Car batteries were so poor the vehicles didn't have the range to finish a 50-mile race: each driver had to hop out of the cockpit mid-session to jump into a second machine. It still hasn't turned a profit. But its founder, politician turned racing supremo Alejandro Agag, always envisaged Formula E as something more. It is an avenue to experiment with new technology and to bring electric racing to life in a way few thought possible.

And so, it has come to pass. Today, each car finishes the race with energy to spare, with a more powerful, energy-dense battery that's capable of being recharged in 45 minutes. The next generation of car will be able to 'flash charge' the battery, recovering 10 per cent of its capacity in 30 seconds. 'We were even talking about wireless charging for the Gen 3 car,' Gorne says. 'And all of that goes back into the public domain. We're going to make *Super Mario*

Kart look quite real. The only people on a par with what we're doing are Formula One or the World Endurance Championship.'

No one doubts Formula E now. It attracts the biggest names in the automobile world: Audi, BMW, Jaguar, Nissan, Mercedes-Benz. Earlier, up in a well-lit suite stuffed with canapés, I'd asked Pascal Zurlinden, Porsche's director of GT factory motorsports, why his team had also joined the paddock. 'For us, this is an information laboratory where you have as much exchange between engines and engineering as you can,' he said. Porsche are here because they know electric cars are the future; already, similar research from the World Endurance Championship has fed into their Taycan supercar. If motorsport is going to help the planet turn a corner, this is going to be a testbed for undoing the damage wrought by exhaust fumes. In fact, it's going to be the only sanctioned option: Formula E has an exclusivity contract with the FIA for single-seater electric racing. No one else can do it.

The cars on track represent the pinnacle of modern engineering, a masterwork of some 11,000 parts at the cutting edge of aerodynamics, simulations, safety and materials. The Envision Virgin Racing cars are identical in appearance (except liveries) to the rest of the paddock. While some championships, such as Formula One, rely on the team to come up with their own designs, interpreting their chassis to set regulations, Formula E teams all use the same Spark-designed chassis. The dramatic trident-like front wing is made of Kevlar and carbon fibre; the chassis more carbon and aluminium; the halo protecting the driver a titanium beam capable of withstanding 125 kiloNewtons of force, the equivalent of 14 Formula E cars piled on top of it. The whole thing, driver included, weighs 800kg – about half an average road car. It is an astonishing piece of design.

Yet the real magic is underneath. Unlike spec series, where all cars are identical and results are solely down to the skill of a team's individuals, Formula E has more than meets the eye. 'The power train of every manufacturer is different,' Gorne says. 'You can have a different MGU, a different inverter, a gearbox, the stiffness of the rear of the car. Everything behind the battery, in the middle of the car, is bespoke.'*

But how did we get here from the Buckeye Bullet? Even at the turn of the twenty-first century, electric cars were a joke; they were milk floats, granny scooters and the cult Sinclair C5. No one would have predicted they'd take over the roads in a matter of decades. What changed?

The answer is in the palm of your hand – and worthy of a Nobel prize.

★ ★ ★

In October 2019, the entire chemistry world cheered in celebration. For decades, there had been numerous calls for one man, John Goodenough, to win the highest award in science. It was looking less likely with each passing year. A Nobel prize could only be awarded to a living scientist and, well … Goodenough was 97 years old. When his name was announced, along with M. Stanley Whittingham and Akira

* If you want to be precise, the teams are only allowed to change the motor, inverter, DC/DC converter, vehicle control unit, final drive, rear suspension, driveshafts, powertrain cooling package, wiring loom and rear subframe. But unless you're super into cars, it doesn't really matter.

Yoshino, it felt a long time coming. Their Nobel prize was for the clever chemistry that powers your laptop, mobile phone and electric car: the lithium-ion battery.

Goodenough is a legend, a man whose life has had more twists and turns than most. An American, he grew up in New England and only turned to science because he was dyslexic and becoming good at maths disguised his difficulty reading. At 17, he was on a hiking holiday in the Norwegian mountains when Nazi Germany invaded Poland; somehow he got back to the US and became a student at Yale, where he had to work to fund his studies. When the US entered the war, Goodenough answered the call, serving as a meteorologist guiding planes protecting Atlantic convoys. Now a radar expert, in 1952 he joined MIT's Lincoln Laboratory, set up by the US Air Force to develop an air-defence system using one of the first computers. Realising the computer's storage was tiny, Goodenough worked with electrical engineers to develop ferrimagnetic cores to store working data and machine code – essentially giving the computer short-term recall. It was the invention of random-access memory, or RAM. Around the same time, Goodenough also formulated the principles of magnetic properties on a chemical level. Today, the Goodenough–Kanamori rules are a cornerstone of modern physics.

The one thing Goodenough couldn't do was teach. By 1980, he had moved to the University of Oxford in the UK and earned a reputation for boring the pants off anyone unlucky enough to end up in his classes. One year, as students began to drop out of the course or skip his lectures, he entered a classroom to find his students had all gone to the pub and had left an audience of cuddly toys in their stead. At least, Goodenough later quipped, the teddy bears managed to stay awake.

But if his abilities as a lecturer were lacking, his prowess in the lab was without question – and he was about to turn his attention to a project that had otherwise been abandoned.

A decade earlier, the oil industry had realised it was going to run out of its cash cow and started looking at alternative energy sources. In 1973, as oil prices quadrupled, Exxon hired Whittingham, a Brit from Nottingham who was working at Stanford University in California, to revisit electric vehicles. He came up with an idea for a new type of battery. As mentioned earlier, the big problem with lead–acid batteries is that they can't store that much energy. Nickel–cadmium, the battery used in the original Buckeye Bullet, wasn't much better. It stored double the amount and produced a measly 1.3V of electric potential per cell. Instead, Whittingham turned his attention to the lightest metal in existence: lithium.

Most of us are familiar with lithium (Li), either through dropping lumps into water and watching it fizz, its use for treating mental health conditions or the song by Nirvana. It was discovered in 1800 by a Brazilian geologist in Sweden who found an unusual rock, which he called petalite. Around 17 years later, the young Swedish researcher Johan August Arfwedson decided to take a closer look at the sample and was able to isolate a light, silvery white metal that was highly reactive. Taking it to his boss, the legendary chemist Jöns Jacob Berzelius, they named the new element after the place it was found: *lithos*, the Greek for 'stone'. Today, we know that lithium is element number three on the periodic table, that the only elements lighter are hydrogen and helium, and that it was first created in the Big Bang, when the universe formed about 14 billion years ago.*

Lithium's lightness means it has a really high energy density, which makes it the perfect candidate for a battery. Whittingham paired a lithium anode with a titanium disulphide (TiS_2) cathode and, with lithium perchlorate in dioxolane as his electrolyte, waited to see what would happen. The design produced 2.5V of electric potential – almost double a nickel–cadmium battery. Better still, it could store almost five times as much energy and 10 times more than a lead–acid battery.

Whittingham's battery also had one major difference over everything that had gone before it. Normally, the reactions in a rechargeable battery eventually cause the electrodes or electrolyte to break down – it's why the battery stops working. In Whittingham's design, the lithium ions wouldn't corrode the titanium disulphide: instead, they ended up between its structural layers 'like putting jam in a sandwich'. This allowed the battery to be charged and discharged thousands of times without losing any performance. It was a process called intercalation.

Unfortunately, Whittingham's design had a major flaw: the batteries caught fire.

Even though the cathode wasn't affected by lithium ions slipping between its gaps, that didn't mean chemical reactions weren't happening. Over time, thin whiskers of lithium metal would build up at the anode inside Whittingham's creations.

---

* Most elements weren't created in the Big Bang and came from nuclear reactions in supernovas or neutron stars colliding. The lithium you see was probably created in stars, too, but it's humbling to know a lump of silvery-white metal has been around since the dawn of time.

Eventually, whiskers would turn into needles, then become spikes that poked through the electrolyte to touch the other side. This handy metal bridge created a direct path for the electrons to get to the cathode – a short circuit. In seconds, temperatures would soar to 500°C, igniting the electrolyte. And if somehow the battery didn't become engulfed in a bonfire of pure red flame, the pressure from the gases trapped inside would turn it into a molten shrapnel grenade.

As you might guess, this was a bit of a problem. After firefighters were repeatedly called to the Exxon labs during the 1970s, they threatened to start charging call-out fees to deal with Whittingham's lithium pyrotechnics. Admitting defeat, Exxon decided to withdraw from battery research entirely.

Now Goodenough entered the fray. He decided to look at Whittingham's design to see if he could improve it and realised titanium disulphide was an awful choice for a cathode. Not only was it too expensive to make, but sulphur is a big heffalump of an atom for this kind of work – number 16 on the periodic table. Instead, he decided to use oxygen – element number eight – which is much smaller and more electronegative, creating an even greater electric potential. After some experimentation, Goodenough made a cathode out of lithium cobalt oxide ($LiCoO_2$), a highly stable structure that allowed even more lithium ions to move in and out. To use Whittingham's analogy, rather than making your jam sandwich out of flimsy, thick bread, you were now squirting jam between hardy, wafer-thin slices of cracker. Goodenough's battery blew Whittingham's away: it had a far greater potential energy than anyone could have imagined.

Goodenough's battery was close to the modern design, but there were still a few chemical tweaks and safety features needed to make it commercially viable. Now it was time for Yoshino's contribution. Working for the Asahi Kasei Corporation, the Japanese chemist kept Goodenough's lithium cobalt oxide cathode but decided to replace the metal anode. Instead, he used something called the rocking-chair principle, where the lithium ions would never have a real 'home' – they'd just move from being sandwiched between two different things. In 1985, Yoshino created an anode made of petroleum coke and found it was perfect. He also decided to add some safety features to stop batteries blowing up. First, he created a thin, porous membrane as a barrier between the anode and cathode. If lithium whiskers caused the battery to overheat, the first thing that melted was the membrane, sealing off the two compartments and stopping the reaction. It was a simple, brilliant idea. And, to make things extra safe, he added a small aluminium cap in the design. If any gases built up, the cap would pop – once again rendering the battery harmless.

In 1991, Sony released Yoshino's creation: a 4V, fully rechargeable battery. Since then, the design of a lithium–ion battery has remained more or less the same, but incremental changes over three decades have near tripled its energy density. High in energy and light in weight, they are the backbone of our modern lives. In 2016, lithium–ion batteries were responsible for about 28 billion Wh of power across the globe; by 2020, it was 174 billion. If you're reading this book on a phone or tablet, congratulations – you're almost certainly doing it thanks to Whittingham, Goodenough and Yoshino.

Lithium–ion batteries still have problems (which we'll come on to later), but they're already completely reshaping how we interact with our world. Today, a modern lithium–ion battery has an energy density of about 250Wh/kg – research is already under way to up this to 500Wh/kg. With none of the heavy moving parts of an internal combustion engine and no need for fuel, electric cars aren't just viable: they're the best option for manufacturers, customers and the planet. They just need a little fine-tuning and an infrastructure ready to use them.

That is where Formula E comes in.

★ ★ ★

The greatest misconception about a pit garage is that it's noisy. You've seen it on TV countless times: a mess of bodies with greasy overalls and clean tools rushing around, yelling over screaming engines as drivers sit, seemingly impassive under the helmets, and wait for the signal to go.

The image you have in your mind's eye is all wrong. A garage is a realm of professional silence and neat, ordered elegance.

There is nothing extraneous. Everything sits in exactly the place it's required to be. No one speaks; as manoeuvres are rehearsed and drilled into muscle memory, there's simply no need. If verbal communication is required, it's done with a quiet, clear direction. Economy is paramount. (Until you win, of course – then you go nuts.)

There are no booths on the pit wall in Valencia. Instead, each garage is divided into quadrants. At the front are two bays, one for each car. At the rear, there's a small storage area and a darkened control room feeding a constant stream of data to the team's mission control. Session over, I cross the pit lane into the Envision Virgin Racing garage, past the lanky form of 'Tower', Robin Frijns' mechanic, who's fiddling with something on the car. Frijns himself passes a moment later, helmet off, hair slicked down with sweat, eyes red and puffy.

'You look tired,' I say, stepping out of the way to allow him to peel out of his flame-retardant suit. He gives me only the briefest glance, shaking his head.

'Not tired. I'm ill.' He gulps back on his water and closes his eyes, sniffing a little. Frijns has a cold but, thanks to the strict anti-doping requirements, he can't reach for a decongestant without running foul of a drugs test. Instead, he has to hurtle around tight corners at twice the national speed limit, simultaneously watching for other cars and managing his on-board systems, without anything to help ease his man flu. (Frijns still topped the session, though, so he's coping fine.*)

I make my way to the control area. Inside is a row of men in matching shirts reviewing the latest telemetry readouts, their eyes scanning over wiggles and shifts on a screen that mean very little to me. While there have been big, sweeping changes in motorsport over the past 100 years, there's no question that the biggest change has come about through this: raw data. While instinct and expertise still have an essential role, it's live feedback that has accelerated technology to new heights.

A Formula E racing car has around 300 sensors on it. Infrared brake sensors for temperature, accelerometers,

* The next day, Robin was fine but I had a sniffle. 'How are you?' I asked. 'Good,' he replied. 'I think I've caught your cold,' I added. 'Good,' he said again with a cheeky grin.

pressure sensors in tyres, detectors to pick up even the slightest movements in suspension over fractions of a second. In Formula One, the electronic control unit – the brains of the car – transmits more than 1.5GB of live data back to its team during a race, sending around 750 million data points for the team to spool through. In total, it'll transmit around 500GB over the course of a race weekend.

This is only a fraction of the data analysis possible; the sport has actually banned telemetry in some areas. This is either to keep costs down, or to prevent anyone having an unfair advantage. For example, Formula E teams aren't allowed to measure wheel-speed movement as this would help them control traction – a driver aid that's strictly forbidden. To sidestep this regulation, some teams took to painting secret markings on their wheels, which were then monitored by trackside cameras. With a little mathematics, the team could work out wheel speed. This simple hack was quickly rumbled by other teams and banned outright, but it illustrates a point: however you can get it, data matters.

Data is so crucial because everything is done on a knife edge; where the distance between first and nowhere can come down to a misplaced decimal point. Each lap, the team's engineers have to run calculations on how much energy they're using, whether they should increase their pace or try and conserve it, and whether their brakes are too hot or tyres too cold. This would be hard enough if the cars only *used up* energy, but some clever tricks with the brakes mean the cars also *recharge* their battery as they race, adding in another variable. And if juggling all of these factors already seems impossible, keep in mind the team has to figure everything out twice (they're running two cars after all) and has less than a minute before the next lap begins.

Track data also feeds into perfecting the car's set-up, although there's no such thing as a 'standard' car configuration to begin with; when you're racing everywhere from the dry heat of Saudi Arabia to the damp dreariness of an east London indoor/outdoor circuit, you have to adjust everything from aerodynamics to suspension. 'You try and find a set-up so that a driver has a baseline,' Gorne says. 'If you changed a car dramatically – gimbals, geometry, springs, dampeners – it's going to feel different for a driver and his confidence will be gone every time he goes into a corner. And you need confidence when you're racing within two millimetres of a wall.'

I turn away from the screens as a figure rises from his place in the booth, taking off his headphones and walking out to meet me with a smile. Sylvain Filippi is managing director for the Envision Virgin Racing team. He leads me out of the garage down to the back of the pits, where tyres are stacked in the sunlight. We find ourselves leaning on a supply crate, standing next to a broken front wing from an earlier accident: the team's other driver, Sam Bird,* got a little closer than 2mm to the track limit and had a brief argument with a chicane. Lean, with a mop of blonde hair above dark-framed glasses, Filippi has been in the sport since its very beginning. After five years at DaimlerChrysler, he'd moved from Paris to the UK to work as a consultant just as electric vehicles were becoming viable. 'It was a complete

* Bird has since moved on to Jaguar Racing.

coincidence,' he says. 'I was working with Nissan to prepare their first electric cars. I got to know the guys at Tesla and I was blown away! I mean, I'm a petrolhead, but an electric engine has extreme performance, instant torque at zero revs. And I loved the smoothness of it, the linearity. You could have a super-high-performance car that was also super nice to drive in a traffic jam. That's unmatched in a petrol engine. So, I thought, "OK, what we need is to go back to our roots and create a racing championship to promote this technology."' Filippi's work became the EVCup and, when Formula E started, he realised there wasn't much point in having two competing championships. 'And here we are, five years later,' he says.

Despite being well established, electric cars still have their critics. One of the most tedious lines is that, because electric cars are just giant batteries and are charged through a grid, they're just as dirty as petrol engines. Does it really matter if the fossil fuel is burnt in the car, rather than in a powerplant? 'The most common thing I get asked, even now, is "Aren't you just making the electricity somewhere else?"' Filippi says, shaking his head. 'Sure, you are. But it's about efficiency. An electric car's engine is 95 per cent efficient, which means you are using three times less energy per mile to travel than a petrol car. It doesn't matter where that energy comes from!'

This is an important point. Even if all of the energy generated to run an electric car originated in a fossil fuel powerplant (and it doesn't), the carbon emissions for the whole chain equates to about 50g of carbon dioxide per kilometre. For comparison, a standard car's CO_2 output is about 150g/km. In fact, driving an electric car is even more efficient than riding a fully packed bus (89g/km per passenger) or train (60g/km per passenger). The 'still using fossil fuels somewhere' argument just doesn't hold water.*

But debunking myths around electric cars is only a fraction of why the series is so vital. 'Formula E is a driver,' Filippi explains. 'If we weren't here, there wouldn't be an incentive, and there wouldn't be a supply chain. Sure, the batteries are the key elements. The more energy we can jam in there, the better. But you have a whole supply chain of motor suppliers, inverters, electronics, all that stuff. Six or seven years ago, if you wanted to procure a motor inverter at a very high voltage, I'm talking about 800V or more, you just couldn't find it. Most road cars are 300–400V. And now – and this a direct benefit of Formula E – you can find major suppliers have very reliable 800V, even 900V systems.'

If you're wondering why a 900V system is better than a 300V equivalent, it's because of a principle called Joule's law. Power is just voltage multiplied by current, so a high-voltage and low-current system could produce the same amount of power as something that's low voltage and high current. The catch is that Joule's law, which throws resistance into the equation, says that as you increase current, you get more heat – it's the reason a light bulb warms up and starts to glow. In a car, it therefore makes sense to use the highest voltage possible, minimising current and reducing unnecessary heat. The downside is that high voltages are dangerous, which is why

* Although we're talking about car emissions here, it's also worth noting that the entire Envision Virgin Racing operation is certified carbon neutral, from using 100 per cent renewable energy and recycling two-thirds of its waste to having a zero-tolerance policy on single-use plastics and red meat. This isn't empty rhetoric – they mean it.

car manufacturers were so hesitant to try them until Formula E came along.

As mentioned earlier, the first Formula E cars weren't as impressive as they are today. But in five years, the cars' battery packs have gone from 133kW to 250kW – with only a 50kg increase in battery weight. 'It's the same ratio with battery capacity,' Filippi says. 'We started at 28kWh and now have 52kWh in roughly the same volume with only a slight weight increase. We've doubled the energy density in four years. All of the manufacturers here will freely admit they're learning about technology. We've moved to silicon carbide for our inverters, which is super-lightweight – we've halved the weight compared with three years ago while generating double the power.'

The obvious question is *how.* 'From everything!' Filippi grins. 'That's what's really fun at this stage!'

★ ★ ★

Innovation in batteries is happening in two main areas. The first is choosing the right materials for the cells. Professor Saiful Islam is a chemist at the University of Oxford who specialises in battery materials and, in 2016, gave the Royal Institution's Christmas lecture on energy. 'Car manufacturers are seeking improvements in energy density, charging, cost, safety and lifetime of the battery,' Islam explains. 'For major breakthroughs, we need new battery materials and a deeper understanding of how they function.'

A lot of this, Islam says, boils down to the exact elements that squeeze into the battery. 'So, the first successful lithium-ion battery was with Goodenough's cathode and Yoshino's anode. But there's been a drive to move away from lithium cobalt oxide because of toxicity, cost and to improve energy density. Cobalt has been partly replaced by nickel and manganese. One of the most well-known acronyms in the area is NMC,* which is nickel, manganese and cobalt, which was initially in a 1:1:1 ratio. Today, car companies are using NMC 6:2:2, cutting down the amount of cobalt. There is even a push to go higher in nickel content, to 8:1:1, but that creates safety challenges. You can also use a bit of aluminium, as they do in Tesla cars. One thing we're exploring in my lab is to increase energy density by stuffing more lithium into the cathode. But there's some way to go before practical applications are possible.'

As mentioned earlier, Formula One hybrid engines use batteries, and their lithium–ion systems have shrunk dramatically. In 2007, the energy recovery system of an F1 car weighed 107kg and was only 39 per cent efficient. Today, it has a maximum weight of 20kg and is 96 per cent efficient. Yet, even though battery technology is improving at an astonishing rate, there are limits – particularly when it comes to charging. 'It's to do with the movement of lithium ions,' Islam explains. 'The charged lithium ions have to move from one side of the battery to the other and in between the electrode layers – which is where the important stuff is happening. We're aiming to speed this rate of ion diffusion up by tweaking the structures of the materials. It's faster to run a hundred metres than it is a marathon, right? So that's

* This abbreviation is not a chemical formula or even the elements' chemical symbols (which are Ni, Mn and Co). It's just what the cool kids say.

why some batteries are getting thinner. It's not just to make them lighter.'

This can cause problems, as there's less gap for the lithium spikes to cross, resulting in short circuits and fires. 'One of the big challenges in lithium–ion batteries is the flammable liquid electrolyte,' Islam says, 'so another avenue being explored is replacing that liquid with a solid. It's a big advantage in terms of safety and you can also increase the energy density by returning to a lithium metal anode. The downside, of course, is that ions move faster in a liquid. Finding a solid that conducts lithium ions just as fast, a 'super-ionic' conductor, while remaining stable would be incredibly important. There are already promising advances in this area.'

Another main downside of lithium is sustainability. Even though it's the 33rd most abundant metal on Earth, the sheer number of devices using it mean that resources could run out. One of the best 'beyond lithium' options is sodium (atomic number 11), which is immediately below lithium in the periodic table and therefore has very similar properties. It's also far easier to get at: three-quarters of the planet is covered in water filled with the stuff in the form of sodium chloride, or salt. This makes it far cheaper to obtain than lithium and a more sustainable resource in the future. But size matters. The problem is that sodium is more than three times heavier than lithium and doesn't have anywhere near the same energy density. There's also an obstacle with the battery's 'jam sandwich' model; because sodium is a much larger atom than lithium, the changes that happen to layers as sodium ions move in and out are far more substantial. The result is that the battery wears out about 10 times faster than a lithium equivalent.

Even so, it doesn't mean these problems won't be solved. 'Chemists are already looking at sodium–ion batteries for large-scale grid storage where the size of the battery doesn't matter as much,' says Islam. 'It's more abundant, there's a lot of it out there and it's a good battery to store energy from renewable sources, such as solar and wind, especially when the sun isn't shining and the wind isn't blowing.'

Although there are sodium–ion batteries for low-cost electric transport – such as e-bikes and scooters – it's unlikely you'll see them in motorsport soon. For now, lithium remains king.

★ ★ ★

Given Formula One embraced hybrid engines a decade ago, you might wonder why we need a solely electric series. The answer is that not all battery cells are equal. By making subtle adjustments to their chemistry, you can change parameters such as their energy density (how much energy the battery stores) and power density (how much energy a battery can release *now*). For example, your mobile phone doesn't need a lot of power to run so you want a higher energy density – which means your battery will last longer. Conversely, if you need a lot of instant juice but don't have to keep your battery running for hours, power density is the way to go.

This means you need different motorsports to overcome separate challenges. A Formula One car can recharge its hybrid engine every lap so it doesn't need to store energy, it needs to unleash it. As such, while F1 engineers have managed to achieve a 12-fold increase in power density in a decade, they have only doubled a battery's energy density. Formula E cars, however, don't have a petrol engine to fall back on: they need enough charge to complete the entire race.* This means their battery research is more focused on energy density, which equates to range. The two sports are tackling different problems and we are reaping benefits from both their answers. And, when it comes to your regular commute, research into making your car go further between charges is far more applicable than improving its performance in a drag race.

The types of changes that bring about these results are far simpler than you might think. Originally, the Formula E battery was built by Williams Advanced Engineering and was made up of 400 cells that looked like large envelopes. Today, a new version is used, made by Lucid Motors from Silicon Valley in the US. Inside an odd, trapezoid shape, Lucid have crammed around 5,000 cells, each about the size of an AA battery. While these smaller batteries can't hold as much charge, they can be arranged in a way to maximise space and cut down weight from the connections. It's this simple modification – the *layout of the cells* – that has led to such rapid gains in the car's range. As Filippi says, the real reason Formula E is able to deliver much higher voltages isn't because of a miraculous breakthrough in chemistry or physics; it's because the sport has created an infrastructure where more people are looking at problems.

Even with a really high voltage, you're still going to get some heat, though – and that means better cooling is essential, too. Every battery has an ideal temperature range: too cold and they're not going to work well, too hot and they'll degrade

* Formula E cars actually have two batteries – one low-voltage system to power electronics and log data and a high-voltage battery to do the real work. We're just focusing on the high-voltage one.

faster. And if they get really hot (about 80–90°C), there could be a 'thermal runaway', leading to a lithium fire. To tackle it, an electric car's batteries are sandwiched between cold plates filled with coolant, which is pumped around constantly to keep everything chilled. The best way to cool electrics (short of something ridiculously expensive like liquid nitrogen) is to use water. Unfortunately, water doesn't go too well with electricity, so Formula E uses an oil-based, non-conductive coolant for extra safety.

A final part of the battery puzzle is software. Formula E cars only get one battery per season and, with 5,000 battery cells all at different temperatures, there's too much for a human to handle. It's therefore down to the car's onboard computer to manage the charge and discharge of each cell throughout the race. At every moment the car is running, the computer is deciding which batteries should be called on, trying to strike a balance between power on track and cell longevity. If a group of cells start to get too hot, the computer just switches them off and uses another. Every year, the software looking after the battery's modules gets better at its job, keeping them in that temperature Goldilocks zone (not too hot, not too cold) by working out when to use them during the race.

These modifications – in design, cooling and software – are just the tip of an engineering iceberg. Far more substantial changes lie beneath. As Filippi mentioned, the sport has started using silicon carbide in its control chips.* This material isn't

* Silicon carbide is also known as carborundum and is probably most famous for the fake Latin phrase *Illegitimi non carborundum*, which supposedly means 'Don't let the bastards grind you down'. Sadly, it's not true. Silicon wasn't discovered until the nineteenth century and by none other than the chap who gave us lithium: Jöns Jacob Berzelius.

new in either electronics, where it's been used for more than 100 years, or in cars, where it's often part of a 'ceramic' brake disc. But before Formula E, no one thought of putting the two together and using it in *car electronics.*

'We're already at levels that five, six years ago you couldn't think about,' Filippi says. 'All we need to do is improve the battery efficiency another big chunk and you'll be past the tipping point. There's no question – you will have an affordable electric car. You'll be able to travel more than three-hundred miles on a charge, and, with faster charging, people won't even notice the difference [compared with filling up at a petrol station]. An electric powertrain is really high performance, but it's very simple, it requires very little maintenance. And we'll get to the magic number – a hundred dollars per kiloWatt-hour. At that point, making an electric car is cheaper than making a petrol one.'

Once past that tipping point, people will realise that electric cars can do things their petrol competitors cannot. As Filippi mentioned earlier, an electric engine is able to operate at very high revs (about 20,000 rpm, much higher than even an F1 car) and can do it instantly without a problem. That means you don't need a gearbox to help your engine stay in its rev sweet spot. Yet this wasn't something designers initially spotted. In the first season of Formula E, the cars all used conventional five-speed gearboxes; it was only through racing that it became apparent there wasn't much point to them. 'The beauty is that almost all electric cars will just be single-speed,' Filippi says. 'That means there are fewer moving parts, so you'll increase simplicity, reliability and availability. It's a beautiful design.'

And all of these fixes – decisions on energy density and power density, battery layout, cooling, the materials to use and the software to run them – aren't staying with Formula E. They're going into the next car you'll buy. 'It's so exciting!' Filippi says, practically bouncing with enthusiasm. 'These aren't changes just for the sake of it. These are absolute, genuine improvements. You're going to see longer ranges, more efficiency and faster charging. There is a direct link with Formula E and road cars coming in the next few years and that hasn't been true of any other motorsport for decades.'

It's an exciting time to be in electric racing. But the battery is only the start. As I mentioned earlier, Formula E cars can recharge during the race. And that isn't down to a new innovation at all – it's thanks to a technology with an origin as old as civilisation itself.

CHAPTER FOUR

Applying the Brakes

You can recycle a nuclear bomb using a dildo. I found out about this unusual science trick towards the end of a very unauthorised tour, deep in the heart of a very secret location, in a country that shall not be named. (Governments, it turns out, are a bit funny when it comes to showing you atomic weapons.) Just getting to the lab itself was a slog. First, there was the ritual of stepping over a raised barrier in the corridor while slipping on disposable shoe covers – a sort of radioactive obstacle course. Next came pass cards, security screens and unmarked elevators with permissive keys to lock out interlopers. Eventually, my guides and I ended up deep in the underground bowels of the facility, safely isolated by thick walls with only big industrial machines for company. It was the room where one of the most powerful countries in the world disposed of its nukes. And there, lying at the heart of this mysterious, secretive and secure space, was an anal vibrator.

It drew the eye pretty much instantly. The interior of a nuclear lab is organised chaos, far messier than most people would imagine, but everything is sanitised and sterile with colours barely ranging beyond monotone. The bubblegum-pink sex toy drew the eye and, being a man of culture, I had to ask why someone had left such a thing next to the world's deadliest munitions.*

---

* I have priors when it comes to this kind of stuff. I once promised a nuclear chemist I'd send her my *Star Wars* poster, autographed by none other than actor

I had two minders with polar views about the device. The scientist rolled her eyes and face-palmed in embarrassment; the engineer giggled like a schoolboy, practically hopping on the spot with glee. As part of the process to sort out his country's nukes, the schoolboy relayed, samples needed to be shaken. Unfortunately, it was hard to find the right piece of kit to do the job. Plutonium is highly radioactive, so the samples are contained behind a row of heavy lead bricks and manipulated through thick gloves. This makes it tricky to shake the test tubes. Anything that even goes close to radiation inevitably gets contaminated, so has to be low-cost and disposable. And, given the extraordinary value of even a single bead of plutonium, any gear used has to be wipe-clean so all the product can be recovered if a spill occurs.

The team had been stuck for a solution. No piece of kit they knew of could stir their samples remotely while filling the other requirements. Then one genius suggested looking beyond the realm of science. Why not get exactly what they needed with a quick trip down to the local sex shop? Somehow, the team convinced their boss to pay for a vibrator and then marched off to the local supplier of intimate pleasure tools. Dildo in hand, they headed back to the lab to try it on a test rig.

Their new purchase worked beautifully.

To this day, the humble vibrator remains the team's prized scientific instrument, happily stimulating plutonium samples and spelling an end to weapons of mass destruction. Sadly, the pink sex toy I saw was not the original: shortly before the

David Prowse, as a thank you for letting me visit. I'd forgotten that Prowse had signed it, 'My lightsaber is bigger than Hayden Christensen's, love Darth Vader!' To this day, there's a Jedi dick joke on the walls of the US nuclear arsenal.

team had put their first toy into use, it broke down. Unable to requisition another dildo from the department, and unwilling to leave a questionable purchase on the family credit card, the researchers decided to send a hapless grad student to get a replacement. So off the student went, into town, into the red light district, into the sex shop, where they slid their receipt over the counter and said: 'I'm really sorry, but after two hours of continuous use my vibrator has broken. Can I have another one?'

The point of this story is simple. Engineers are problem-solvers who spend their lives trying to come up with quirky ideas that are more creative than their rivals. Ever since the nuke incident, I've called such quirks an 'engineer's dildo' – taking something and translating it to a different aspect of life where it takes on an entirely new purpose. When you add such creativity to the hyper-competitive world of motorsport – a place that attracts eccentrics and where rules are open to interpretation – you create a frothing stew of people constantly trying to come up with clever ideas that bend, twist or downright break the paradigm.

Why do these rules exist? Paramount is safety. The shape and pace of change is also adjusted to try and keep a level playing field, such as limiting teams to a specific amount of research in one area, or curtailing a device that could give an unfair advantage. Perhaps more importantly for us, they are also made to stop sports drifting away from reality. This does not mean governing bodies, such as the FIA, always get it right. At the 1976 Spanish Grand Prix, master outside-the-box thinker Ken Tyrrell unleashed Project 34 – a six-wheeled Formula One car, with four small, 10-inch tyres at the front and two 'normal' F1 tyres at the back. The principle behind the idea was simple: more tyres, more contact patches with the road, providing greater traction when turning. Although the idea was laughed at (Sir Jackie Stewart choked on his drink when he heard about the news), the six-wheeled Tyrrell driven by Patrick Depailler qualified third. By the time the car reached Sweden, future world champion Jody Scheckter led the Tyrrells to a one-two finish. Sadly, the car design had huge problems – the dinky tyres wore out far more quickly, so the car needed more pit stops, and the driver needed little portholes to be able to see them and monitor wear. Eventually, the concept was abandoned, but it didn't mean it was forgotten. In the early 1980s, numerous teams, including Williams and Ferrari, looked at six-wheeled racers even if they went against existing rules. Eventually, the FIA mandated that Formula One cars had no more or fewer than four wheels and the idea was shelved for good. If Williams had been allowed to carry on down that design path, who knows what we might be driving today.

There have also been numerous F1 innovations that have been granted a life in the world's toughest motorsport, undergone a rapid arms race and vanished back into the world of cars. During the 1980s, anti-lock braking systems (ABS) were trialled before being banned as they were deemed too useful for drivers. So too was McLaren's 'brake steer' or 'fiddle brake', an extra peddle that allowed the driver to only brake on one side of the car enabling them to take corners faster. Traction control and four-wheel drive have also seen similar fates. And while none of these inventions, except the fiddle brake, originated in racing, the names associated with them are known for speed. The first electric ABS brake system was on Concorde; traction control was developed by Buick, which won the first race at Indianapolis Motor Speedway; and the first four-wheel drive car was built by Ferdinand Porsche. Take apart any modern car and it's virtually impossible to find something that isn't tied in some way to going fast.

And that can lead to some unlikely connections.

★ ★ ★

If you've ever wondered why modern cities expanded so rapidly – outwards, downwards and upwards – it's thanks to an American called Frank J. Sprague. Born in 1857, Sprague was a maths prodigy who intended to go into the army but ended up accidentally taking the four-day naval entrance exam instead. He obtained the highest score of his intake and quickly developed an interest in naval electronics. By 1881, the 24-year-old Ensign Sprague had designed a new type of dynamo and the first electric bell system for a US Navy ship. Quickly, Sprague's talent was recognised and, in 1883, he was poached by Thomas Edison. The two men didn't get on: Sprague wanted to work on motors while Edison wanted him to focus on light bulbs. The next year, Sprague left Edison and set up the Sprague Electric Railway & Motor Company. It's probably a good thing he did: Edison was an asshole who had a penchant for stealing credit from his employees.

Sprague's company soon began to electrify the world. He improved the way electric trams collected power from overhead lines and built tram systems for Richmond and Boston. By the end of the decade, more than a hundred electric railways had been set up, heralding the start of rapid transit systems. The inventor designed train junctions and signals, and created the electric 'third rail' system for Grand Central Station in New York. In 1897, his railway in Boston went underground – the first electric subway system in America – while his cars rattled above the traffic on elevated tracks across Chicago's South Side. In 1907, his subway cars became the rolling stock of the Paris Métro; although their classic box-shaped carriages no longer thunder through the city's Art Nouveau stations, many survive in working order to this day.

Lifts also benefited from Sprague's fertile mind. In the nineteenth century, these were usually powered by steam or hydraulics, but Sprague realised electricity would not only speed up the lift's travel, but also take up less space than the alternatives. In 1894, the Sprague Electric Elevator Company opened its doors and, for the first time, users could pick the floor they wanted at the push of a button. By the 1920s, Sprague had even come up with a way to run local and express lifts in the same shaft. Without him, modern skyscrapers wouldn't be what they are today.

While his successes paved the path to modern city living, another of Sprague's innovations was also proving its worth. The principle behind braking is simple: you've got a spinning wheel and if you jam something against it that causes friction, it slows it down. As mentioned earlier, energy is never lost, only transferred. When something brakes, the kinetic energy from its moving wheel is usually converted via friction into heat. Sprague wondered if, instead of losing all that energy, he could somehow put it back into his trains.

Sprague arranged it so the traction motors on his trains would flip during braking. Rather than providing energy, they would gather it, turning into generators and slowing the wheels at the same time. When you put this energy into a resistor and it's dispersed as heat, it's called dynamic braking; this is the kind of thing you see on powerful locomotives, which would otherwise rip their brakes apart. But Sprague went one better. He arranged for the energy to be diverted to his power supply and regenerative braking was born.

There are loads of different ways to create a regenerative braking system and multiple ways to store the energy. In effect, though, they all do the same thing – act as energy recyclers – so that instead of losing it as heat, you can use it to power the vehicle. It's an easy way to make something more efficient and can have a huge impact on both performance and the environment. For example, between 2004 and 2007, the Delhi Metro system estimated regenerative braking alone reduced its CO_2 emissions by more than 90,000 tonnes.

Small wonder, then, that motorsport has been using multiple versions of regenerative braking for more than a decade – and even borrowing ideas older than recorded history to do it.

* * *

Late autumn on the Côte d'Azur. A pleasant ripple of breeze cutting off the balmy Mediterranean. A gentle sweat on my brow from the heat. Blue waters, pastel-shaded towns and glorious excess. I've slipped on a train from Italian Ventimiglia, through the French enclave of Menton, into the futuristic subterranean station of what is, according to the UN, the wealthiest country per capita in the world.

Monaco.

The very name sparks glamour. Monaco, the sun-kissed playground of the elite, former home of Grace Kelly, the place where James Bond comes to chance his luck at the Monte Carlo Casino's baccarat tables. Stepping out of the station's tunnels upon one of its moving walkways, I can't help but feel a little of that stardust rub off. Stretching my legs, I decide to relive a giddy, childhood dream and make my way down to the port, past expensive cafes and designer outlets, to the most famous racetrack in the world.

Racers have been drawn to Monaco since 1929, when William Grover-Williams, gentleman and spy, beat all comers in his Bugatti 35B. The Monte Carlo Grand Prix circuit is one of the few race courses you can drive yourself: rushing up towards Casino Square; around the spinning, steep hairpins of Mirabeau as you descend towards the sea; swooping through the Larvotto Tunnel then out into the dazzling sunlight of the harbour's chicanes; around the swimming pool and past La Rascasse …

But skipping about town pretending to be a Formula One driver isn't why I'm here. Across the principality, on the other side of the Rock of Monaco and its palace battlements, lies Fontvieille. Born from the sea on reclaimed land between the 1970s and 1990s, this is the newest stretch of the tiny country. I'm here to accept an invitation I received back in Ohio: a chance to visit the headquarters of Venturi.

The offices are brand new, climate controlled to perfection despite the humid air outside. In the showroom-style foyer sit their feats of engineering. There's the Voxan Wattman, the fastest electric motorbike in the world. It's flanked by a concept car that looks to have emerged straight from the set of *Bladerunner.* And, as a centrepiece, the Buckeye Bullet 3, fresh from its trip over the Atlantic. Now with its skin on, the craft looks like an angry, electrically powered missile as it points out of the showroom's windows – directly at the Ferrari dealership across the street. Electric cars are coming, whether you're ready or not.

It's here you'll find another of the Buckeye Bullet's alumni, someone who has traded the Utah desert's merciless salt pan for the track. Delphine Biscaye is Venturi's Formula E team manager, handling all of its day-to-day needs. We missed each other in Valencia but that's hardly surprising – even in the relatively relaxed atmosphere of a test track, there's a lot going on.

'The chaos is part of the fun,' Biscaye grins, describing a typical day at the races. 'It's being a test lab, trying out and developing new technologies, and it's marketing, too, trying to sell electric cars, but we are still a sport. Part of my job includes driver debriefs between sessions, where they'll say where they got understeer, where they braked going into a corner, the lifting points. They talk, we listen, take notes. And then you've got the technical debrief with the engineers as well as the drivers. That's where we look at the data and say, "OK, the brakes were too hot there, the tyres too cold here." Then we look at the driving and compare their data with other teams. After, we adapt the car, we discuss the setup, we pass that to the mechanics. We have a quick word with the driver to say how the system's changed, what they should feel. And at the end of the race, you've got the same debriefs about what was good, what went wrong, as well as strategy. Did we take the attack mode at the right time? Should we have gone later? What happened compared with other drivers?'

Once all that's dealt with, there's still the fault list – everything that went wrong – for the team to analyse. And that goes beyond damaged parts; it could be down to as much as whether there was a slight leak, or if the battery charge was off by a fraction of a per cent. And after that? 'Uh, it's terrible,' Biscaye says. 'When you've been doing that for a whole week, and you're tired as well, you've still got to pack up! Then we arrive back here, we'll review the full list, assign all the faults to people and find solutions. We'll do a proper race analysis, the drivers will come and we'll all watch the race together again, and try and see what we missed. And then … ' she laughs, 'and then, it's time for full analysis and preparation for the next one. And it all begins again.' You can understand why making time for a journalist isn't all that easy.

Originally from a sleepy village in France, Biscaye stumbled into motorsport almost by accident. Her family weren't Grand Prix fans and, as a mechanical engineering student, she'd applied to the French Institute of Advanced Mechanics without any intention of going into racing. 'When I arrived at my interview, the teachers asked what I wanted to do and I had no idea,' she recalls. 'But everyone else was talking about motorsport, so I just said that. The teachers told me I'd better pick something else, that there were very few places and not many women. They weren't being sexist, it was just a fact. And that's when I said, "OK, then I'm *definitely* going into motorsport!"'

Determined to beat the odds, she began applying to Formula One and Le Mans teams and found a job at Williams. And it's here that Biscaye helped connect Sprague's trains with the future of motoring: she was asked to help develop Williams' take on a kinetic energy recovery system, or KERS.

KERS is the best-known form of regenerative braking in motorsport. The principle is simple: take the energy from the brakes and use it to give your car a boost when the driver needs it, such as coming out of a corner. It was first trialled by McLaren in the 1990s, but its hydraulic system was ruled illegal. Instead, KERS entered Formula One in 2009. While most of the devices functioned in more or less the same way (putting the energy into a battery), Williams relied on a far older concept, a flywheel, to get things working.

Flywheels – big, heavy spinny things – have been around since ancient times, when they were used in spindles and potter's wheels. While they're hard to get turning, once they go they take a lot of effort to stop. They're often used in cars to even out torque from an engine so that you get a nice, consistent rate of revolutions, rather than a huge variation depending on where the cylinders are in their stroke cycle. More importantly for our story, in the Industrial Revolution, engineers realised they were essentially acting as stores of kinetic energy, using a principle called the conservation of angular momentum. In the late eighteenth century, Scottish inventor and all-around genius, James Watt, developed huge flywheels that could store massive amounts of energy, revolutionising factory work and giving Victorian engineers something to pose in front of. Today, flywheels have been perfected to the point that you can pull out 90 per cent of the energy stored in them for use with other things.* New York State, for example, uses a huge flywheel plant to help manage its electrical grid. Rather than risk a blackout caused by a power spike, it stores about 20MW of energy inside 200 flywheels, each weighing more than a ton and spinning at 16,000rpm. The flywheels are capable of meeting 10 per cent of the state's power demands, so if a surge occurs – for example,

* Why only 90 per cent? Because no matter how well-oiled your flywheel is, it still experiences friction. As long as friction exists, a flywheel has to stop eventually; otherwise it's a perpetual-motion machine, which breaks the first law of thermodynamics (that energy can't be created or destroyed, only converted).

if everyone in New York City suddenly decided to turn their kettle on – the energy is there waiting to be used.

The Williams KERS system took this concept, shrunk it to fit in a car and applied it to the brakes. 'When I was with Williams, we were working on the first prototype,' Biscaye says. 'The idea was to create an energy source that would go back into the car. When you brake, you make the flywheel turn and store energy.' This is the simple version: the reality is a little more complicated. The Williams' regenerative brakes turned the motor into a generator, which then fed an electric circuit in the centre of the flywheel. Normally, this wouldn't do much, but Williams coated their carbon flywheel with magnetic particles. This turned the wheel into a big magnet, which spun around at 50,000rpm as it was charged. When needed, the polarity of the voltage was flipped, and the momentum in the wheel created a current that fed into a motor to give the car a little extra push. It was simple, compact, lightweight and didn't lose efficiency. 'Williams Advanced Engineering did the flywheel, and the motor was done by the Williams team, which is what I worked on,' Biscaye recalls. 'It was fun! We were trying to be the most innovative. You could have these crazy ideas and we'd try and model them in 3D, make it, test it and see if it works.'

Sadly, the Williams flywheel never made it into F1. In 2010, refuelling was banned, and the space where the flywheel sat was needed for the enlarged fuel cell. In fact, KERS in general had a rocky start in the competition, with the whole paddock agreeing not to use KERS systems in 2010 at all. But that doesn't mean Biscaye and the other engineers worked in vain. In 2012, Audi's hybrid cars won an incredible 1–2–3 at the 24 Hours of Le Mans using the Williams flywheel. And, in 2014, a fleet of 500 London buses were fitted with the system too, hoping to cut down their emissions by one fifth as they drove around the UK capital. Not bad for tech stolen from Ancient Egyptian pot-makers.

By the time Williams' flywheel was aiding London commuters, Formula One had moved to hybrid engines and KERS had been fully embraced by the paddock. Today, a more advanced version is used called the MGU-K (Motor Generator Unit – Kinetic), which makes up an F1 car's energy recovery system along with the exhaust-based MGU-H. The amount of energy that can be recovered is capped (2MJ per lap), although it's still enough juice to restart the car if it suffers a power failure – as Sergio Pérez showed at the 2021 Bahrain Grand Prix.

But, as mentioned, there are several ways to run a regenerative braking system. And Formula E, being all electric, can ensure the energy can go straight back into the main battery.

'We have regen levels of 250kW,' Sylvain Filippi says. 'We can regenerate as much energy as we put out, which is gigantic! To get that kind of regen on a road car you'd have to put on the handbrake and stop as hard as you can, it's a huge amount of deceleration.' Even though it's far beyond the levels a road car would use, it's still an important part of the learning process. 'It's about calibration,' Filippi explains. 'We are proving that you can generate such regen that in the future you'll hardly need brakes. You'll still have them for safety, but 90 per cent of your deceleration will be without [wearing out] the brake pads. You'll get more energy back into the battery, more range, and you'll probably never have to change your discs and pads for the whole life of the vehicle.'

Formula E teams have also looked beyond regenerative brakes, seeking other ways to save up energy. 'Take Nissan in

2019,' says Biscaye. While the rulebook limited the teams' power output to 250kW from a battery, there were no rules saying the energy could only go to one motor. So, instead of a single powertrain, Nissan decided to have two of them, using some smart, sneaky engineering to save up energy and use it when needed. 'Having that second storage system meant that, in the end, you've got more power than just the battery output. It was legal but it was a grey area, so now there are rules that say you can't have a second energy storage system, you can't have capacitors and all that sort of stuff. Even so, every year, engineers are going to try and find another grey area.'

This is why motorsport is so fun: no matter what branch you're in, someone will always be looking at the tech regulations, trying to squeeze out an advantage by reading between the lines. Although, when it comes to Formula E, it's surprising that the part on show probably gets the least amount of development. 'There's no point in developing aerodynamics or the chassis,' Filippi says. 'It just creates an arms race where you spend more money to generate downforce. It's fun, it makes racing cars go faster – but it really has no application whatsoever to the real world.'

Or does it?

CHAPTER FIVE

The Last Airbenders

You can drive a Formula One car along a ceiling. I know this because once, moments before beginning an interview with an engineer, he'd been running the numbers to prove it. 'You'd be fine,' he told me. 'You just need enough downforce, first to overcome the weight of the car being pulled down by gravity and then to keep the tyres pushed into contact with the ceiling. Even the slightest imperfection would be enough to make you fall off, but you'd be fine as long as you don't hit a crack, bump or light fitting. Of course, that's assuming you have an engine that can run when it's upside down ... '

Unsurprisingly, nobody's been dumb enough to try it and you're not going to see Sir Lewis Hamilton bobbing inverted along the Larvotto Tunnel any time soon. One person has done similar things in the past, though. Willem Toet is a former aerodynamicist at Benetton, Ferrari and BAR, and a consultant at Sauber and Alfa Romeo Racing. He knows a thing or two about air. Once, he took the wheel of a Formula One car and held it against the ceiling. 'If we tape up the holes,' he told his audience, 'put in a little tube and connect it to an old vacuum cleaner, we should be able to take a wheel, place it on a ceiling and, once we've turned the vacuum cleaner on [to suck out the air], with any luck it will stay there.' Sure enough, with the aid of vacuum alone, Toet managed to make the wheel stick. To prove his point even further, he then jumped up, grabbed the wheel's rim and hung off it. The wheel, glued to the ceiling by nothing but the power of aerodynamics alone, took his weight.

'Anyone who's studied aerodynamics knows it has to start with mathematics,' Toet explains. He's right: if you really want to study the area, you need to do some pretty tricky sums. You have to get your head around air-flow principles, such as Bernoulli equations that show how pressure decreases as flow increases; Navier–Stokes equations to work out the flow of viscous Newtonian fluids; or even the Venturi effect, which means that fast-moving fluids suck slower moving fluids around them. (And air, or any other gas, is a fluid.) But don't panic just yet. To grasp the basics of aerodynamics, you just need to know that it comes down to one thing: swimming through an invisible pool. As you move through water, the water also moves around you, filling in the void you've left behind. The same thing is happening as you pass through air.

'In the end, we are working with molecules and we're moving them out of our way,' Toet says. 'In about a centimetre, there are nearly 3 billion molecules in a line. That's about 26 million-million-million ($26x10^{18}$) molecules in a cubic centimetre. That's why we can think of it as a continuous fluid – it's just a lot of individual molecules jiggling around.'

This jiggling was discovered by a completely different branch of science. In 1827, botanist James Brown was watching pollen grains in water under a microscope when he noticed they jinked and darted about: a process he called Brownian motion. This is because the particles in a fluid aren't static; they constantly move at random, jolting and bumping into each other. This fluctuating, ever-changing pattern – like a molecular mosh pit at a rock concert – is air's natural state.

And when you put something solid in the way, comprised of molecules that are tightly bound together and stuck in place (such as, for example, a car), the molecules in the air have no option but to wiggle around it.

It's easier for a fluid to move around something nice and pointy than something long and flat – it's why it's much more fun to dive into a swimming pool than to do a belly-flop. In racing, this air resistance, acting against the motion of the vehicle, is called drag. The problem for racers is that drag increases with the *square* of velocity but the power needed to overcome it increases with the *cube* of velocity. In other words, if you double your speed, you get four times as much drag and you need eight times as much power to overcome it. This is why there's a limit to how fast a car can go: the faster you move, the more molecules you're having to push out of the way and the more you get slowed down. Eventually, this goes beyond the power output of your engine.

Drag is measured by something called the drag coefficient. The coefficient of a human, for example, is 1.0–1.3 depending on how bulky they are; whereas the Eiffel Tower, being harder for wind to get past, is about 2.0. A normal car's drag coefficient is about 0.3, whereas the wing of a fighter jet, which wants as little drag as possible, is about 0.02. And a Formula One car? Surprisingly, depending on its set-up, it's between 0.7 and 1.0 – around three times greater than a road car. This is because of all the wings an F1 racer has to maintain its grip on the road. While it's important, drag isn't all a Formula One team cares about.

As air molecules rush over the car, they will exert pressure where they hit. And this is where engineers can get cunning. The various laws of physics mean that as air flow speeds up, pressure reduces. The molecules struggle to get past the obstacle (*i.e.* the car) and bunch up, but then seemingly speed up and spread out once they've got a clear path in front of them. So, by directing the air flow with wings, dips and curves, you can make the air help you out. If you look at the wing of an aeroplane, for example, you'll see it's curved at the top, flat at the bottom. This makes the air flow faster over the top of the wing than those continuing underneath. This means there's lower air pressure above the wing, generating lift. If the wing were shaped the other way, the flow would be faster below the wing, creating downforce.*

By the 1960s, racing car designers realised they could use downforce to push the tyres harder into the track, creating more grip. And, by tweaking the wing position, you can change the set-up of your car. Less downforce makes you faster in a straight; more downforce makes you able to corner more easily. Away from lift and downforce, you can also use the air to your advantage in other ways – such as sucking it into your engine to cool down overheating parts. And we're still not quite done. As the car moves, it leaves a hole for the air to fill. This doesn't happen in a nice, smooth way; it churns up the air, causing wild, chaotic turbulence. Going back to the swimming analogy, as you move through a pool, the water goes from a neat wave to a swirling mess as you pass. This wake can cause pressure changes behind the car, resulting in drag. Controlling the air you've just passed can be as important as controlling the air coming at you.

---

* You're not actually changing the velocity of an individual air molecule here. Rather, you're changing the orientation of its movement; any air that's going in a direction you don't want gets shoved out the way. This results in an *average* velocity change.

All of these factors require a host of calculations to figure out, and still only scratch the surface of what car engineers need to worry about. When air passes through the engine, for example, it needs to leave as exhaust. This means hotter air is mixing with the cooler air – exactly the same temperature variations that cause turbulence that affects aircraft. And if this wasn't mind-boggling enough, this isn't happening in a single dimension. It's happening in 3D, affected by temperature, humidity, air pressure and wind speed, as well as the car's weight, weight distribution and velocity – none of which stays constant during a race. And, speaking of races, remember that you're sharing the track with 19 other cars, all of which are messing up the air too.

The point is that aerodynamics are so important, they're something that every motorsport has to worry about. And that includes the ones with a reputation for not worrying about much at all.

★ ★ ★

In Raleigh, North Carolina, the local breakfast delicacy is fried chicken and waffles. And if this strikes you as a little odd, the state capital's history museum is even stranger. Given North Carolina is the birthplace of powered flight, you'd expect the replica of the Wright Brothers' *Wright Flyer*, the world's first aeroplane, to be given pride of place.* It's there,

* The Wright Brothers were from Ohio but launched their plane in North Carolina. This has led to the two states engaging in a weird argument over who discovered aeroplanes. In 2001, each state got its own quarter dollar coin: NC was 'First Flight', while Ohio was 'Birthplace of Aviation Pioneers'. Honestly, this is the kind of pointless science war I live for, but sadly it's got nothing to do with motorsport.

hanging from the rafters, but comes a distant second to a car parked in the lobby: a big, bruising, black Chevrolet, its aerodynamics even more advanced – if more subtle – than the flyer suspended above. It's the car that Dale Earnhardt finally, on his 20th attempt, drove to victory in the 1998 Daytona 500. It was the win that cemented his legend as one of the greatest NASCAR drivers of all time.

NASCAR is a religion in North Carolina and Earnhardt its patron deity. While further west the phrase 'The Man in Black' might mean Johnny Cash or Stephen King's demonic antagonist, in NC it'll always mean Earnhardt. Aggressive, relentless, Earnhardt was said to have the ability to 'see air', using the laws of aerodynamics to do things other drivers couldn't contemplate, let alone achieve. Formula One had Ayrton Senna. NASCAR had Dale Earnhardt.

NASCAR is a motorsport with an even more colourful history than the madcap, organised races of Europe. In 1919, the Eighteenth Amendment introduced Prohibition to the US, banning the sale and consumption of alcohol. Just about everyone ignored it; even the President, Warren G. Harding, turned the White House into a speakeasy, famously drinking his way through his term in office with prostitutes dancing on the tables.* This created a demand for Appalachian bootleggers, who modified their cars for speed and handling as they sought to evade the police while delivering their moonshine whiskey. After the repeal of Prohibition in 1933, the bootleggers needed

---

* Harding's sexual adventures are the stuff of legend. In addition to drinking, he spent hours playing poker, having 'meetings' with his mistress in the closet (the Secret Service standing watch outside) and writing smutty letters on behalf of his penis, Jerry. Again, this has nothing to do with motorsport.

a new source of money and took to racing their tricked-out cars. By 1948, NASCAR was born. 'NASCAR is the American South,' explains physicist Diandra Leslie-Pelecky, who wrote *The Physics of NASCAR: The Science Behind the Speed*. 'Baseball came out of the civil war; football and basketball came out of colleges; and NASCAR came out of good ol' boys and liquor runnin'. NASCAR was never about technology; it's about loud cars with lots of horsepower. And, in a way, that's what's held it back. A single family owns and runs NASCAR. The first two generations ran it like a benevolent dictatorship and the third nearly ran it into the ground. They're just now realising that their future depends on moving not only the car, but the entire culture, into the twenty-first century. It's frustrating for those of us trying to defend NASCAR that driver Richard Petty continues to say that women have no business in racing and that it took until 2020 for NASCAR to ban the Confederate flag.'

Despite this 'good ol' boy' persona – or maybe because of it – NASCAR is the most popular spectator sport in the US, attracting 75 million fans to watch its races and boasting 17 of the top 20 most-attended sports events in the country each year. Its tracks, such as Daytona and Talladega, have passed into legend, even beyond the US. You might never have heard of NASCAR's first commissioner, Erwin 'Cannonball' Baker, but you've probably heard of the Cannonball Run named in his honour. And in 2006, Will Ferrell's NASCAR comedy *Talladega Nights: The Ballad of Ricky Bobby* created the catchphrase 'Shake 'n' Bake', since used by F1 rivals Sir Lewis Hamilton and Sebastian Vettel. (The title of the movie is a joke, by the way: Talladega Superspeedway doesn't have floodlights, so there's no racing at night.)

Despite its popularity in the southern US, NASCAR gets a lot of derision from other motorsports because of its layout. Although seven of the tracks on the calendar are road courses, such as you'll see in Formula One, many of NASCAR's most famous tracks are banked ovals. This has led to critics describing it as a sport that basically boils down to who can turn left the fastest. It's a reputation Eric Jacuzzi, senior director of aerodynamics, simulation and design at NASCAR, is all too familiar with – but it hides a lot of science going on underneath. 'I mean, yeah, it's a fair statement,' he says. 'You're in an oval. But at places like Charlotte the banking is 24 degrees, you're entering a corner at 190mph. It's a *rollercoaster ride* left. If you take a regular car and go run the Daytona road course, you'd be in the wall with blown brakes in two laps.'

While NASCAR is technically a stock series – the cars being raced are, in theory, the same you could buy from a dealer – the reality is they are nothing of the sort. These cars are modified and tricked out to perfection, well beyond the specs seen on the road. The aerodynamics of even turning left have seen some pretty crazy things and highlights why it's so important to think of a car's travel in three dimensions. 'So, the ability to generate sideforce and yawing movement [the direction the nose of the car is pointing] is huge,' Jacuzzi explains. 'Back in the old days, 10–15 years ago, they'd build cars asymmetrically. So, the front of the car would be symmetrical, but the back of the car would actually be permanently bent over to the right. That meant you could get the actual body of the car, relative to the wheels, to move out so they could take the corner skewed sideways. They figured out you could generate all this sideforce. Sure, it makes the car crap down the straightway. But when it gets to a corner it's bent in eight degrees, but the tyres are only at three degrees.' What the cars lost on the straights, they made up for tenfold in being able to take corners faster than their rivals. Sadly, NASCAR eventually realised the teams were going too far in bending their chassis to maximise the use of sideforce. 'The cars looked like kidney beans, it was embarrassing,' says Leslie-Pelecky. 'The fans started calling it the "Twisted Sister",* which didn't make manufacturers happy at all.'

Of course, some motorsports want exactly the opposite effect the NASCAR designers were aiming for. Land-speed racers, such as the Buckeye Bullet, are built to stick in a straight line. 'Little tweaks make gigantic differences,' says Buckeye Bullet team manager David Cooke. 'Drag reduction is huge to be able to go faster, but so is stability. The key to land-speed vehicles lies in darts thrown at a dartboard. There's a big heavy metal weight at the front, so it's very front-heavy, and giant fins at the back, right? So, the centre of mass is in the front. The centre of pressure, the aerodynamic centre of the dart, is in the middle, where you put your finger. This means that when a dart starts to pivot in the wind it pushes back – there's a tremendous amount of stability. So, when you throw a dart, it releases and flies straight. Basically, we want to do the same thing with a land-speed car: use aerodynamics to keep it in a straight line.'

This isn't as easy as it seems. The car needs a little instability in case the driver needs to steer out of the way of an object, or adjust for crosswind: get it 'too right' and your car is basically on railroad tracks. Working out a car's weight

* I guess the fans were trying to tell NASCAR that 'We're Not Gonna Take It'…

balance is relatively straightforward but pinpointing its aerodynamic centre of pressure is far harder – you need a computer to do it. And land-speed records also have to factor in tyre–ground interactions, too, which can play havoc on aerodynamics. That's hard enough on tarmac; the Buckeye Bullet team need to do it on wet, clumping, naturally craggy salt.

Land-speed teams and NASCAR engineers have their aerodynamic challenges, then. But both will freely admit there's one sport above all others when it comes to the mastery of air: Formula One.

★ ★ ★

Formula One's battles with aerodynamics have changed the course of the sport. In the early 1970s, Lotus dominated races thanks to its wedge-shaped Lotus 72, a design light years ahead of its rivals. Unfortunately, aerodynamics were still not fully understood and Lotus took the decision to remove the car's wings. The result was – as Cooke outlined previously – huge instability problems, which ended with tragic results (which we'll discuss later). The wings returned and the wins continued, eventually resulting in three constructors' championship titles.

Even weirder attempts at aerodynamic prowess have been trialled in Formula One, too. Perhaps the strangest was the Brabham BT46B: also known as the 'fan car'. By sticking a huge fan on the car's rear, supposedly for cooling to make it legal, Bernie Ecclestone's Brabham team was able to generate a huge amount of downforce. The result was incredible: despite being uncomfortable to ride, driver Niki Lauda found it had amazing grip going around corners. His rivals hated it. Lotus' Mario Andretti likened it to 'a bloody great vacuum cleaner' that 'was throwing up a lot of dirt and shit', while Jody Scheckter attached a desk fan and bin lid to the back of his racer with a message proclaiming 'Bernie sucks'. Lauda's response was blunt: 'If you don't like it, you should overtake or fuck off!' Despite the driver's defence of his car, it only ran for one race – Lauda using it to win the 1978 Swedish Grand Prix – before Ecclestone withdrew the car voluntarily to keep everyone happy.

This is just a taste of the aero tricks the sport has seen. 'Formula One is really a competition between engineers,' says John Hearns, who was a lead high-performance computing engineer at McLaren for five years. 'It's the only sport for engineers that exists. But there are no shirkers in Formula One. I'm Glaswegian, and there's a saying we have in Glasgow: keeping someone out the way of the buses. There are many companies in the world that keep people on just to keep 'em off the streets. In Formula One, there's no room for that. They're gone.'

Hearns joined McLaren in 2008, spending six years as an engineer in the organisation's aero team. In his first season, McLaren won the drivers' championship, with Sir Lewis Hamilton – then only 23 years old – collecting his first title. 'That was a high,' Hearns says. 'Then, at the beginning of the next season, we weren't competitive at all, to say the least. You'd go into the office and you could cut the atmosphere with a knife.'

Most people don't realise the true size of the Formula One paddock. The crew that goes to each race is the tip of an iceberg: often a mere 10 per cent of a complete Formula One team. Getting the cars ready and competitive is the undertaking of thousands. And there's no mystery where the teams place their focus. 'In a Formula One team, you have mechanical engineers, structural engineers, electronics engineers. But about half of the team is aero engineering,' Hearns explains. '*Half* of it. And these are all guys who could work on fighter planes – they've all got top-class aeronautical engineering degrees and PhDs. I've known people who could have gone on to design aircraft but they felt that they'd only see the things they'd design fly once in their career. In Formula One, they see a new thing they've helped to create appear on the car every two weeks. That's what attracts people.'

Why does it take so many people to calculate aerodynamics? Because it's the one area of the sport where huge amounts of time can be gained without massive overhauls, where a slight change of wing angle could win you a fraction of a second. In something like NASCAR, this doesn't really matter. In Formula One, it can be the difference between pole position and crashing out of qualifying in the first round. 'The aerodynamics of a Formula One car make up about 25 seconds a lap,' Toet says. 'And that's really hard to find elsewhere.' For reference, a typical F1 lap is about 1 minute 20 seconds, so aerodynamics are responsible for a full *third* of the car's time.

Aerodynamics calculations are run constantly. And they cover all aspects of a racing car, even down to where the broadcaster positions its cameras. 'There are sections of a car, they look like little wings, that are actually camera positions,' Hearns says. 'They're supposed to be aerodynamically neutral. But where you place them can make a difference in your performance.'

The real challenge with a Formula One car, Hearns explains, isn't to do with the air coming in but rather what happens once it's passing over the car. 'It's calculating turbulence rather than smooth flow. That takes serious smarts. Formula One cars are open-wheel racers, they have an open cockpit, too. Turbulence modelling is very difficult. If we had mudguards on the wheels like a conventional car, it'd be far easier. Instead, you have to model how the wheels turn, down to the contact patch – the part of the wheel that is on the tarmac – and the yaw angles as the car is going around the corner. We even have to model how two cars pass each other.'

Perhaps the greatest example of how this control of turbulence can influence a race happened in 2009, when Brawn GP – an entirely new team formed from a management buyout of Honda – raced with a secret aerodynamic weapon. 'They started racing and were just beating everyone in practice,' Hearns remembers. 'We were wondering what the heck was going on. And then we finally saw the back of their car.'

The 2009 championship had seen a huge shake-up in aerodynamic rules. The FIA had realised that all the little tricks teams had used were creating dirty air that prevented overtaking. The 2009 cars were therefore limited to what they could do, and the days of hundreds of small wings, bumps and channels were over.* Instead, the engineers had started to look at the exact wording of the rules to see where they could

---

* The irony is that these rules never worked. The idea was for teams to only have 60 per cent of the downforce they had previously. While that happened, the teams worked out their cars would be better if they were wider. This prevented overtaking – and thus made all the changes pointless from the start.

gain an advantage, and the Brawn team had stumbled on a loophole. The FIA rules said there couldn't be any holes for aerodynamics in two important parts of the car, the step plane and reference plane. But it didn't say anything about the *transition* between those two parts. This allowed them to make their diffuser – the part that accelerates air flow under the car, creating low pressure and maximising downforce – even bigger than the legal limit. Suddenly, the top teams were running a 175mm diffuser, while Brawn had the equivalent of a 300mm device delivering downforce.

'One of the guys in my team figured out what it was immediately, and just went, "Oh my God!"' Hearns recalls. 'Basically, the regulation said that if you looked up from underneath you're not supposed to see the bottom of the car through the bodywork. And what they'd done was, basically, put in Venetian blinds. You couldn't see the bottom, but there was still a gap.'

Brawn's 'double diffuser' didn't just make them more competitive. While all the other teams tried to catch up, Brawn racked up victory after victory, giving them such a lead they went on to win the drivers' and constructors' championship titles. A single piece of aerodynamics had made all the difference.*

A year later at McLaren, a brilliant aero engineer had a concept for something that was equally controversial in the sport: the F-duct. A small intake duct peeping out of the body,

---

* Brawn were not alone in working out the double diffuser; Williams and Toyota also came up with similar inventions. Brawn's was simply better. Nor is it completely fair to put Brawn's victory solely down to the double diffuser; Jenson Button was outstanding that year.

the F-duct was simple in principle. Formula One cars weren't allowed moveable aerodynamic parts, but McLaren realised if they could somehow make their rear wing flatter on the straight without a mechanical control, they'd cut down on drag and get a higher top speed. Thinking outside the box, the team took inspiration from the military. 'We looked at US Navy fighter jets,' Hearns says. 'An F-4 has small wings, which are great for manoeuvrability, but when they came in to land they didn't have enough lift. So, they had a switch, which would bleed off air from the engine and blow it over the wing to give a boost. Now, in F1, you're not allowed movable parts on the car. So, we didn't have any. We got the driver to use the back of their hand instead.'

The F-duct was normally harmless: a little crossover that routed air through the car and out the back. The trick was that one of the duct's openings was through the cockpit and the driver could block it. This was the sporting equivalent of the Dutch boy sticking his finger in the dyke. In an instant, the driver had created what's called a fluid switch, resulting in a change in air pressure that rerouted air flow to under the wing, causing it to flatten and slash the car's drag. Now, McLaren could have the wing up for tight corners and, with a flick of the wrist, flat for the straights. 'In the end we told [the other teams] what we were doing,' Hearns says. 'Because everyone could see the drivers on the telly, waving their hands about. It wasn't secret for very long.' Today, the F-duct is banned, replaced by the drag reduction system (DRS), a way to mechanically flatten the car's wing that has specific rules on when and how it can be used.

At this point, you're probably wondering what this (very simplistic) primer on aerodynamics has to do with green technology. Well, it's because the work that's gone into perfecting air flow – from NASCAR to Formula One – has led us to a point where we can use these tricks to slash our energy bills and better plan our cities.

All you need to do is take the next step and ask a computer to do the maths for you.

CHAPTER SIX

Going with the Flow

In the early 2000s, Rob Rowsell was looking for a job. Young, fresh-faced and almost at the end of a master's degree in engineering, Rowsell was part of a new breed of engineer, one born after the advent of the computer age. While at the University of Manchester, Rowsell had developed his skills in a technique called computational fluid dynamics, or CFD, that used processors to model air flow. Scouring the job ads, he spotted one for a small company based in Brackley, Northamptonshire, called Advantage CFD.

'I thought they were a workplace safety company!' Rowsell laughs. 'When I arrived, I thought it was a bit odd they had all these pictures of Formula One cars all over their office. Unbeknownst to me, they were actually the CFD department for the BAR Formula One team. I worked there for six months then, when I graduated, for another six months. They seemed to like me.'

Entirely accidentally, Rowsell had stumbled into the early origins of computer modelling for sport. 'CFDs had been around for quite a long time, but initially it was a bit fringe,' Rowsell says. 'It was only about 20 years ago, as computing power got better and better, that the usefulness of the technique improved. Twenty years ago, it wasn't accurate enough for aerodynamicists to really believe in it, but it gave them insights when you did wind tunnel tests. There was still a lot of guesswork and supposition, though, and [aerodynamics] was down to imagination and experience.

Today, the accuracy and raw numbers have got better and better and better.'

The simplest way to envisage CFD is as a geometric mesh, Rowsell explains. 'It works by basically breaking down the 2D or 3D representation of whatever you're trying to model into little blocks. Inside each of those blocks, or cells, you solve a set of equations. You then pass the answers to those equations back into the cell itself, doing another loop, and pass the equations on to neighbouring cells. There are lots of different types of equations you can do, too; either you can make it faster to solve or you can be more accurate, which takes longer.'

Since the advent of aerodynamics, the gold standard of testing was (and still is until you get out on track) the wind tunnel. This lets you see exactly how your aerodynamics play out thanks to a giant-ass fan blowing air over your car or model. With the aid of smoke, dye or even bits of string attached to your subject, you can physically see what's happening to the flow. Wind tunnels also let you see minute changes in air pressure, measured by placing your vehicle on an incredibly accurate balance; the weight of the car doesn't change, but by detecting slight variations in the balance's readings, you can tell what's happening with lift and downforce on the car. (Kind of like when you tilt a bit on bathroom scales, trying to make the number more to your liking.)

The problem is that massive tubes blowing gale-force winds aren't cheap. Wind tunnels are expensive to build, run and maintain, which means time spent in one is at a premium. Most small-scale teams couldn't even dream of owning their own and, even when the tunnels are available (NASCAR's Research and Development Center has two full-scale tunnels for teams to use, for example), usually they're only there as a final touch to prove the calculations are right.

This is why Advantage CFD existed in the first place. It had started in 1997, founded by IndyCar supremos Reynard Motorsport* and BAR to become the leading CFD consultants to the race industry. As one of the smallest teams in the paddock, BAR had no hope of competing with the likes of McLaren on even terms. And, unlike the big teams, they didn't have access to their own wind tunnel. Instead, they had embraced computer models and quickly expanded their portfolio beyond F1. While half of the team worked on modelling a Formula One car's aerodynamics, the other half took on contracts for gas masks, wind turbines, inhalers or 'anything that might want some fluid to flow over it'. This wasn't just to generate money: it was to feed back into the design of the model and make the car faster. 'Diverse things improve your overall CFD capability and knowledge,' Rowsell says. 'You cross-fertilise and it improves the F1 effort.'

Advantage was an early leader in CFD. Rather than just deliver raw data, it showed its clients computer-generated imagery to help them understand what was going on. Soon, it was working with Ferrari, Aston Martin and Honda as well – the latter eventually buying out the BAR team. In 2007, the party was over: Advantage was folded entirely into the Honda Racing F1 team.† But it had kick-started an industry.

---

* Strictly speaking, Reynard raced at the time in Championship Auto Racing Teams (CART).

† As race fans will know, this means Advantage became part of the team that would eventually win nine (at the time of writing) constructors' championships, first as Brawn and then as Mercedes.

Rowsell wasn't there. After university, he took some time off to go travelling, and in 2004 came back to competing job offers. He could have gone to Kraft; he could have stayed with Advantage CFD and worked on designing different types of nozzles for home coffee machines. Instead, he opted to join a former client of Advantage, who wanted to bring their CFD effort in house.

They were a small company owned by a maverick designer with a robot dog.

★ ★ ★

The man who had recruited Rowsell was Nick Wirth. In 2001, Wirth had launched RoboDog at the Institute of Mechanical Engineers in London, boasting what was then the largest autonomous robot with legs in the world. It was the size of a Labrador; could beg, go for walkies or roll on its back; had the strength to carry a small child and weighed about 12kg. Sadly, Wirth's robot dog never really made a dent in the market (it cost £20,000 at launch). I only mention it because it's what comes to mind every time I think of Wirth. I wonder if, somewhere, that robot dog is getting battery-charged pats for being a good boy.

Even without a cyberhound, Wirth's career is astonishing. In 1988, he entered motorsport, aged just 21, as apprentice to legendary designer Adrian Newey. At 23, he went into business with future FIA president Max Mosley, creating the racing company Simtek. By 1994, Wirth was in charge of his own race team: Simtek Grand Prix. It was, on reflection, a disaster. In the first race of the 1994 season, its lead car (driven by David Brabham) qualified 26th, while Roland Ratzenberger didn't manage to qualify at all. Although Brabham finished 12th in the race, he was still last: everyone behind him had retired. In the second race, Brabham's car was out with an electrical fault, while Ratzenberger limped home to 11th – again, last of all the cars running. The third race was the infamous San Marino Grand Prix. We'll come to that later.

Simtek folded in 1995 from a combination of lack of sponsorship, crippling debts and potential backers withdrawing after being affected by an earthquake. With it, Simtek Research went bankrupt. Wirth dusted himself off and started again. That led to Wirth Research – and the hunt for someone to run its CFD programme.

Rowsell remembers his start at Wirth being far from the supposed glamour of Formula One. 'I built our first CFD computer in my bedroom, working at home,' he says. 'Then another, then a few more. My housemate started grumbling about the number of computers we had! In the end, I had about 16 PCs hooked up in a little cluster.'

With his homebrew machines, Rowsell found himself working on a project to win one of the triple crowns of motorsport: the Indianapolis 500. Along with the Monaco Grand Prix (which, traditionally, falls on the same day) and the 24 Hours of Le Mans, it's considered the pinnacle of the business. To date, only one man, Graham Hill, has managed to win all three.

The Indy 500 race takes place at Indianapolis Motor Speedway. Known as the 'brickyard' because, in 1909, the racing surface was paved in masonry, today the track is coated in tarmac save for a single yard of bricks at the end. The rules are simple: 500 miles, or 200 laps of the track. It sounds a long way, but keep in mind the lap record is a mere 37.895

seconds. Each year, 33 drivers line up to shoot for glory, driving for the right to chug the traditional celebratory bottle of milk, cheered on by a jubilant crowd of 350,000 spectators. It's also one of the deadliest races in the world. Since it first ran in 1911, 58 people (including 11 spectators or track personnel) have been killed.*

'I was working on Honda's IndyCar effort,' Rowsell recalls. 'The Indy 500 was, by far, the most important race of the season. Basically, you've got to develop a car that's as low drag as possible, with just enough downforce so that the driver can go around the two bends at the end flat out, no braking, no lifting off. That's the challenge – you're doing the whole lap with your foot down. If you have even slightly more downforce than you need to get round, then you've also got too much drag, meaning someone is going to be quicker or burn less fuel. If you don't have enough downforce, you'll gradually slide off and go into the wall. So, even though it looks like a very simple race, you've got to get the car tuned perfectly. You've also got to account for the tyres warming up, too, which means balancing front and rear downforce. It takes really high precision.'

Honda had invested in wind tunnels but was relying on Rowsell's bedroom set-up to give the team more insight. Combined, it was enough; despite having to drive through a severe thunderstorm that would later turn into a tornado, the Honda of Buddy Rice won the Indy 500. Two years later, Wirth's engineers were working on the American Le Mans

---

*There's another unsavoury aspect to the Indy 500 in that women were effectively banned from participating for decades; so much so, women reporters weren't even allowed in the pits until 1971. Stay classy, IndyCar.

series and realised they didn't need the wind tunnels any more. 'Nick sat down and ran through all of the areas of the car to develop, and we realised that, for every single area, the aerodynamicists, previously very strong wind tunnel guys, all preferred CFD. That was the tipping point.'

The next project for Wirth was Manor Motorsport.* One of the new teams to enter Formula One in the late 2000s, Manor had been attracted to the sport with promises of a budget cap – a promise that was never realised. Suddenly, they were faced with running a team on a shoestring compared with their rivals and there was simply no space for a wind tunnel programme costing around £1 a second. 'The amount of money allocated to Wirth Research to actually design and build a car was relatively small,' Rowsell recalls. 'I think we worked out that, by the time we'd built the first wind tunnel model to start developing from, we'd have used up the entire aero programme's budget. So, at that point we realised we had to become the first people to design and deliver a Formula One car entirely in CFD.'

By now, the biggest challenge for CFD wasn't the detail of the models – a Formula One car's CFD model can have upwards of 500 million cells as part of its grid – but the speed of processing. While CFD can be done on normal computers, usually it requires more power than the average home desktop. You are, after all, modelling millions of individual molecules to paint a mosaic of how air will react when it hits your car. Wirth's programme to design a Formula One car, running 24 hours a day, seven days a week, used cores

* The team raced as Virgin Racing and later Marussia Virgin Racing, but were originally Manor Grand Prix.

running around 40 trillion calculations a second. (The team actually had almost double this capability but, like all things, the FIA put a cap on how much development could happen.) To put this into context, Data from *Star Trek: The Next Generation* used to claim his speed was 60 trillion calculations a second, while a 2022 top-of-the-range modern computer console runs at about 12 trillion calculations a second. Wirth's system was *fast*.

The car the team designed, the Virgin VR-01, struggled thanks to its limited design budget and reliability issues. Yet at the other end of the grid, the larger teams had embraced CFD as well. 'I remember one season we needed a new supercomputer,' Hearns says. His role inside McLaren's aero team was as an HPC, or high-performance computing, engineer. 'One of the team's partners specified and assembled the system in Chippewa Falls in the US.* It was phenomenal. The racks were flown over on a jumbo jet, offloaded, and we put it in a space underneath [team principal] Ron Dennis' office. In the high-performance computing industry, these machines can take literally months, maybe even a year, to actually be tested so you can measure their performance. We installed it on Monday and were running production aero jobs on Thursday. We had it going in four days. It was unknown to do it that fast, just unknown.' What impact did the supercomputer have? 'Well, we designed a new front wing on it,' Hearns laughs. 'The next race, Hammy won, so ... '

In the 10 years since, CFD supercomputers have become an industry standard – and make up the majority of aerodynamic

---

* I always thought this was just a made-up place Jack is from in *Titanic*, but it turns out it's in Wisconsin. Same thing, really.

design. Today, CFD caps have been removed, giving Formula One teams a huge opportunity to develop their race programmes. Processing speeds have also increased dramatically. In 2021, NASCAR's CFD programme, run at the Ohio Supercomputer Center, used 23,292 cores with *each* containing 128GB of memory. 'When I started doing CFD for teams, seven or eight years ago,' NASCAR's Eric Jacuzzi recalls, 'I would do three cases a week, maybe, because it would take so long. We'd do perhaps two hundred cases a year. Last year, we probably ran three thousand simulations. NASCAR's next-gen car [set to debut in 2022] was mostly designed with CFD, we did very minimal wind tunnel testing.'

This produces a *lot* of data. As mentioned earlier, an F1 car produces around 500GB of data during a race weekend. But a Formula One team's factory will produce around 350TB of car data annually – mostly from CFD, but also from wind tunnels and test rigs. That's roughly 424 times the hard drive of a PlayStation 5.

Yet while CFD has taken over car design, it's far from its only use when it comes to sport. Just ask Great Britain's Olympic team.

★ ★ ★

Most of the top Formula One teams have an applied engineering branch. Separate from the team, yet integral to the overall company, these branches focus on taking F1 know-how and transforming it into real-world solutions. Applied engineering teams work on a host of different areas, meaning that, even if you've never seen a Formula One race, chances are you've seen their work in action.

It's a world Matthew Williams knows well. An aerodynamics engineer with a PhD from City, University of London, he joined McLaren in 2006 as a CFD specialist before becoming a customer engineering project manager for McLaren Applied Technologies. Williams has done everything: from taking the lead on aerodynamics of the racing versions of McLaren's road cars through to working on other sports entirely. At the 2014 Winter Olympics in Sochi, Williams was part of the team that designed Great Britain's bobsleigh. Team GB claimed bronze, ending a 50-year medal drought in the sport.* It wasn't the first or last time Williams helped Great British winter athletes, either: the skeleton bobs used by Amy Williams and Lizzy Yarnold to win gold at three consecutive Winter Olympics also have McLaren's fingerprints on them.

'I originally got involved providing support to the UK's Olympic cycling programme,' Williams says. 'Then winter sports, not only the sled design but how the flow interacts with the athletes on the bob itself, that kind of stuff.' Having built a pedigree supporting national federation programmes, McLaren then took its expertise to partner with commercial brands, most notably Specialized Bicycle Components. 'We went straight into winning the Tour de France, green jerseys, world championships, things like that.'

Things like that. Williams almost shrugs off his involvement, underplaying what was arguably the greatest British sporting success story of the twenty-first century – one that few people realise had a helping hand from Formula One. 'Our involvement

---

* On the day, the British team finished fifth. They were only awarded their bronze medals in 2019, when two Russian teams ahead of them were disqualified for doping.

in cycling probably changed the way the cycling industry works,' Williams says. 'Now, everybody does computer modelling, wind tunnel testing, validating performance – not just experts saying it'll save you 30 seconds in a race, but being able to quantify exactly what each performance change means.'

The McLaren bike revolution started with one of the biggest names in the sport: the 'Manx Missile', sprint specialist Mark Cavendish. In a bike sprint race, the final few metres are where the race is won or lost. Designers, therefore, put all of their focus on the bike's stiffness, ensuring maximum power transfer. McLaren took the existing team bike and made its frame 15 per cent lighter, but still managed to make its bottom bracket – where the rider transfers power to the crank – 11 per cent stiffer than before. It didn't take long for the design to be validated; it won its first outing, ridden by Matthew Goss, at the 2011 Milan–San Remo (one of the most prestigious one-day events in the sport). Later that year, Cavendish used the design as he won the points classification of the Tour de France. The same year he was appointed an MBE and named BBC Sports Personality of the Year. Today, he's widely recognised as the greatest bike sprinter of all time.

Cycling. Sailing. Rowing. Canoeing. Name any British Olympic medal winner of the past decade who uses a piece of equipment and chances are Formula One gave an assist. And F1's involvement isn't limited to aerodynamics; it also helps with performance management. As discussed earlier, an F1 car's sensors are designed to be small and to measure changes over fractions of a second. This means they can also be stuck on bikes or sleds to give feedback and show athletes exactly what's happening in their initial launch or final dash to the line.

And, like Rowsell's, Williams' work isn't just focused on sports. 'I've been involved in fluid prediction for the oil and gas industry, through to modelling work for medical devices, domestic products and many different forms of transport. I mean, I've probably been involved in everything that could have fluid floating around it except spaceships. I haven't done any spaceships yet.' He stops to think.

'Well. Not for McLaren ... '*

At McLaren's rival, Williams Advanced Engineering, this extra-curricular work has resulted in some very unlikely partnerships. Using CFD, Williams worked with the UK Ministry of Defence to develop its Biological Surveillance and Collector System (BSCS), a device designed to scoop up particles as they pass to alert troops of an incoming biological or chemical attack. The BSCS – built on Formula One tech – entered service with the Royal Air Force in 2017.

But while uses in the Olympics or on the battlefield grab headlines, you can also see Formula One's aerodynamics in action much closer to home. In fact, we reap its benefits every time we go to the supermarket.

The path between Formula One and the supermarket freezer isn't straight. In fact, it veers off in an unexpected direction entirely: long-distance transport. 'Wirth had been doing some projects with [haulage firm] Eddie Stobart on their trucks,' Rowsell explains, 'trying to make them more aerodynamically efficient. We'd modified around 400 of them, creating kits to improve their air flow. It saved each vehicle

---

*That's right – Formula One teams work in space, too. For example, All American Racers (which competed in three Formula One seasons as Anglo American Racers) produces the landing arms for SpaceX's Falcon 9.

around 5 per cent fuel burn, which when it comes to a truck is massive.'

The success with Stobart prompted Wirth to look at other vehicle modifications designed to reduce fuel consumption and noticed that Marks & Spencer trucks had aero kits on them, too. 'We went to them and told them that, based on our simulations, we didn't think their lorries with the striking and visual curved roof trailers were working as promised in real-world conditions. The head of sustainability there said, "Well, actually, our biggest issue is refrigeration in stores."'

Think back to the last time you were in the supermarket. Most of the refrigeration units don't have doors on them: they're open-fronted cabinets, designed for convenience, where people can grab their food and move on. While it's easy for customers, it's terrible for the environment, as it's the equivalent of leaving an entire aisle of fridge doors wide open. 'The cold air spills out and is replaced by warm air that goes into the fridge and needs cooling down,' Rowsell explains. 'They're really inefficient. Typically, 30–60 per cent of a supermarket's energy is taken up by refrigeration.' If Wirth could somehow stop the cold air spilling out, it could save Marks & Spencer millions. More importantly, it could have a national impact. Around 3 per cent of the UK's energy use is from supermarkets or convenience stores; if Wirth could stop the air loss, they could make a huge difference to the UK's carbon footprint and environmental legacy.

If you cut a supermarket fridge unit open, you'd see it works by two circuits. The first is a closed loop containing a refrigerant – such as Thomas Midgley Jr's CFCs. Essentially, the refrigerant chills the food as it evaporates and turns into a gas, gives off all the heat it has absorbed at the back of the fridge, cools back into liquid and then goes around for another pass. It's actually a little more complicated (involving pressure changes and the laws of thermodynamics), but that's the rough idea. The part Wirth was concerned with was the second circuit: the air blowing around the fridge. This was also supposed to go around in a circle. If you look in a supermarket fridge, you'll notice there's a grill at the bottom. Here, air gets sucked in, blown through fans and coiled past the refrigerant, which makes it even colder. These streams of cold air come down from holes at the top of the fridge cabinet in a kind of waterfall effect – called the air curtain – which keeps all of the items at the right temperature.

The problem is that the air curtain is easily broken. Almost immediately, the stream will hit shelves, items and customers' hands that are reaching in. This causes the cold air to fall out of the fridge unit and escape. 'If you look at the CFD model, you can see it gets very turbulent and very messy,' Rowsell says. 'It's mixing with the warm air from the aisle. Warm air is going in and cold air is coming out and going into the aisle around customers' feet.'

If you've ever wondered why it feels chilly walking around in sandals down a supermarket aisle, now you know it's thanks to faulty air curtains. And it was here that Wirth could make a difference. 'We realised that if we could recapture the airflow and get it tight again, nice and smooth, then we could reduce mixing in the air curtain,' Rowsell continues. 'Think of a waterfall. In nature, it isn't just falling all the way down. It cascades as it hits rocks and a lot of air is pulled in, which churns it up and makes it white and foamy at the bottom. Whereas if you look at an architectural waterfall, say the Trevi Fountain in Rome, the water stays clear. That's because each time it spills over a lip, it's recaptured and the stream remains nice, tight and smooth. That's what we wanted.'

Wirth's solution was simple: stick the equivalent of a Formula One car's rear wing on each shelf. 'We developed what we called EcoBlades,' Rowsell says. 'It's just a pair of blades with a slightly converging duct in between. The air curtain flow falls in, gets tightened and straightened out. As it falls to the next shelf and starts getting chaotic again, it gets caught by the next blade [and straightened out once more]. Just fitting these to a typical supermarket's shelves saves about 25 per cent of the store's total electric bill.'

In addition to lowering energy costs, this simple change cuts down huge amounts of food waste, as it keeps the produce at the correct temperature. And, by preventing the air spilling out on to customers' feet, it also makes shopping more pleasant. 'It also cuts down the store's heating bill during winter,' Rowsell says, 'because they're not cooling the store accidentally by spilling air out of their fridges. And during summer, fridges often can't cope with the extra drain from the warmer temperatures and will start switching off. With the EcoBlades fitted, they're not sucking in as much warm air from the aisle, so they can keep running all summer as well.'

The EcoBlades are already installed in Waitrose and Morrisons supermarkets across the UK. And similar developments are seen elsewhere. Sainsbury's, for example, introduced Williams Advanced Engineering's 'Aerofoil' – a similar innovation yet again taken straight from a Formula One car. In 2015, the supermarket trialled it in 50 of their stores. The result was similar to Wirth's design: Sainsbury's have claimed their fridge power use had been cut by 15 per cent, equating to saving more than 8,700 tonnes of CO_2 per year. Five years later, more than 1 million aerofoils have been installed in the UK with almost every major supermarket using similar blade designs on their fridge shelves to keep their food fresh and their carbon footprint down.

CFD has taken aerodynamics out of the wind tunnel and into our daily lives. And there's one more advantage to computer-based aerodynamics: you can apply them to something truly massive.

★ ★ ★

Skyscraper design is hard. For the architects and engineers responsible for the project, there's an awful lot to take into account. You've got to consider building materials, entrances and exits, temperatures, power and utilities, and weird X factors that probably don't enter into most peoples' minds. In 2013, for example, the 37-storey London office block at 20 Fenchurch Street *melted someone's car.* The building, dubbed the 'Walkie Talkie' because of its convex shape, was alleged to act like a reflective mirror, focusing the Sun's rays down on to vehicles parked nearby and causing their bodywork to buckle. The only way to solve the problem was to fit the building with a giant sunshade, which prevented it acting as a makeshift death ray.

One of the biggest of these X factors is wind. Taipei 101, for example, was the tallest building in the world from 2004 to 2009, dominating the skyline of northern Taiwan. At 509m tall, the wind that howls around the building's roof is powerful enough to cause the whole structure to sway, which meant its architects had to take action. Corner shapes were used to cut down wind vortexes and crosswinds;

high-strength concrete was added up to level 90 to provide additional stability; and a giant, 18ft wide, 726-ton ball, known as a tuned mass damper, was stuck inside the top few floors, suspended by four pairs of steel cable. This ball is tuned to the resonant frequency of the building, acting as a pendulum to sway in the opposite direction if the building starts to move. As it does so, it passes any kinetic energy on to a series of dampers that dissipate the energy into heat (similar to how dynamic brakes work in Chapter Four). With the help of its huge golden ball, Taipei 101 can resist typhoon winds of up to 134mph and is certified to withstand the strongest earthquakes ever recorded. It's an incredible feat of engineering and well worth a visit if you're in town.*

Smaller skyscrapers generally aren't at risk of being blown over, but they can affect the wind at ground level. At my university, one of the taller buildings was renowned for its wind spill, the structure acting as a giant fin that funnelled airflow directly across its main entrance. By my final year, my friends and I would start walking towards the entrance from about 3m to the left, confident that, by the time the wind blew us sideways, we'd walk straight into the main hall without having to adjust. Once, we even had an umbrella race, opening up our brollies along a makeshift starting line and letting them go to see which one would fly away the furthest.

Modern planning authorities often require data to confirm that new high-rise buildings won't create these wind spill

---

* Mass dampers were used as part of the suspension system of Renault F1's cars in 2005 and 2006. Reputed to reduce lap times by an astonishing 0.3 seconds, the damper helped Fernando Alonso claim back-to-back drivers' championships. Sadly, dampers were ruled illegal by the FIA.

effects. Traditionally, such evidence was hard to come by, as skyscrapers don't exactly fit into wind tunnels. But with CFD, that doesn't matter: modelling a 30-storey building is no different from modelling a 30cm wing tip.

'Actually, it is,' Rowsell corrects. 'We ended up with models that were about twice the computational size of an F1 car. We've been looking at architecture since 2010. We started working with [architect firm] Foster + Partners to model air flow around buildings. Sometimes this was on the outside of buildings in the cityscape and sometimes inside. And, sometimes, we've done both.'

Let's stick with the outside first. Just down from the Walkie Talkie, one of the newest skyscrapers in London is 22 Bishopsgate, a 62-storey, 278m building that's the second-tallest in the City of London. The architects realised that the building would cause major downwash and create a miniature vortex for pedestrians walking past, so Wirth was asked to correct its wind spill. In three days, the team were able to draw up 25 design variants, each with different vanes and baffles that could channel the wind somewhere else. Ultimately, Wirth settled on a stack of six curved baffles, placed 8m above pavement level, which removed the building's downwash entirely. The design was then passed back to the architects to incorporate into their overall aesthetic.

CFD can also place a building in its real-world environment. In a built-up metropolis such as London, Chicago or Hong Kong, skyscrapers aren't just surrounded by smaller buildings, but other massive fin-like structures playing havoc with the wind. This is where a CFD technique called transient analysis, or unsteady flow analysis, can come in handy. By placing the building in its geographic context, CFD can improve air flow, stopping downwash while dissipating noxious gases and pollutants, and also calculate the pressures a building's cladding will face. Wind-driven rain models can even work out where water's going to collect after a storm, allowing planners to place drains and downpipes in the right spots to prevent flooding. Puddles can be a thing of the past.

The power of CFD to aid planners inside a building is just as revolutionary. Staying in London, Wirth helped design Bloomberg's new billion-pound headquarters next to Mansion House. It's a building that's designed to breathe. 'Normally, new office blocks and buildings are fitted out with air conditioning for the whole year,' Rowsell explains. 'Using CFD, we helped the architects develop schemes that would pull in air from the outside and move through the floor plan on each storey of the building. The air gathers in an atrium, then circulates through the vents and roof, all naturally. There are no fans, no kind of mechanical forced ventilation at all.'

This uses a principle called the 'stack effect'. The difference between air density inside and outside a building (thanks to temperature and moisture) results in a difference in air buoyancy. It's an old principle, one used to ventilate mine shafts or, more tragically, one often seen in devastating fires. By working out how to obtain a stack effect with CFD, Wirth's engineers designed a safe ventilation system that would keep the Bloomberg HQ's air fresh and comfortable. 'Obviously, you still need air conditioning, or other ways to cool air, for the hottest periods of the year,' Rowsell says. 'But even so you're going to save a lot of energy.'

This is only one of the tricks Wirth employed to govern Bloomberg London's airflow. Inside the building, CO_2 sensors determine how many people are occupying each area, and a computer opens and closes vents to channel the airflow accordingly. Overall, smarter airflow is estimated to reduce the office's CO_2 emissions by 300 tonnes a year. It's part of the reason Bloomberg London is ranked as the most sustainable major office building in the world by the Building Research Establishment.

If this sounds impressive, in 2017 Wirth took on an even more ambitious project, when it helped to design the airflow for Apple's new $5 billion HQ in Cupertino, California – a huge hollow ring wider than an ocean liner that houses 12,000 employees. This time, Rowsell's team had to plan for the weird weather of the San Francisco Bay area and the impact of being surrounded by a small forest. In the end, the team modelled the wind effect for each of the 7,600 trees nearby to get the building's airflow just right.

The lessons Wirth learned from this project have fed back into your weekly grocery shop. 'From our work with Apple's HQ we've combined it with the refrigeration stuff,' Rowsell says. 'So, we're now looking at supermarkets and helping them to model their entire store. There's a big problem with wind infiltration, when cold air comes through their open front door and cools the store down, meaning they have to have the heating on. Or, in hotter climates, the other way around – hot air comes in and you need to use too much air conditioning.'

This problem has led to Wirth's latest innovation, a technology called AirDoor, co-funded by the UK Government's Innovate UK. The team places a CFD-designed archway in front of a supermarket entrance. This arch contains an array of sensors that detect which way the wind is blowing. It then adjusts its own in-built air channels, creating an opposing, self-generating breeze to act in counterpoint. The result is a hidden barrier of air: an invisible door to stop wind infiltration and reduce energy wasted on heating and cooling the building. Already, it's been installed in 11 Waitrose stores, as well as two Morrisons supermarkets.

It's incredible to think that the future of our cityscapes will stem from 16 PCs scattered around Rowsell's bedroom to win the Indy 500. But the potential impact of computer simulation on our lives goes much further. Already, scientists have started to model an entire human body. And when you trace the technology back to its roots, you discover it starts with David Coulthard playing video games.

CHAPTER SEVEN

Virtually There

On 21 August 2005, the inaugural Turkish Grand Prix took place at Istanbul Park in Tuzla, not far from the storied banks of the Bosporus. It was a sunny day, with track temperatures in excess of 45°C, but that hadn't deterred 100,000 spectators from coming to see the first race on the new surface. Designed by renowned architect Hermann Tilke, the racetrack ran anti-clockwise through 14 turns and over four different ground levels. It featured a corkscrew first turn and a tricky turn eight that swept rapidly through four apexes. It was fast, ruthless and relentless. Formula One supremo Bernie Ecclestone boasted it was the best circuit in the world.

The 2005 Formula One championship had been a close-fought battle between Renault and McLaren, pushing Ferrari, the team that had dominated the sport for five years, into a distant third. As the motorsport media descended on Istanbul, everyone wondered which team would be able to adapt to the track fastest. All had seen the layout from the design blueprints, but none of the drivers had ever done a lap. The general consensus was that the sport's new star, Fernando Alonso, would dominate.

And yet, when it came to practice, it was the McLarens that crushed the opposition. While others learnt the racing line, the McLaren drivers just seemed to *know*. In the first session, the McLaren of test driver Pedro de la Rosa was lapping 1.4

seconds faster than Alonso.* In qualifying Kimi Räikkönen was comfortably ahead of both Renaults and he'd go on to win the race. His team mate, Juan Pablo Montoya, narrowly missed out on second place after an accident with a back marker, but still set the fastest lap of the race – 1:24.77 – a record that still stands.

While other teams were baffled by the McLarens' instant familiarity with the track, their success wasn't down to psychic powers or better instincts than their rivals. They had just been driving its virtual doppelgänger every day for weeks.

'Back then, we were the only team with a simulator,' remembers the system's designer Caroline Hargrove. 'Before Turkey, we were always told that drivers could just learn tracks quickly, blah, blah, blah. But here it was clear: the simulator helped the drivers learn the track. They were faster than everyone from the word go. I remember Juan Pablo coming on the radio saying, "This is just like the simulator!" At that point I was like, "Yes! We've made a big step!" It was a massive win for us.'

Hargrove is a mathematician and mechanical engineer. She grew up in Montreal and her childhood was filled with the sonorous, echoing sounds from the nearby Grand Prix track. 'I remember growing up and there was a French Canadian, Gilles Villeneuve, driving for Ferrari. That was a big thing for us. It was like Ayrton Senna in Brazil: every French Canadian was a Formula One fan when Villeneuve drove. So, I was always a motorsport fan, but working in Formula One was never a career ambition.' Instead, after studying at Queens

---

* The bottom six teams of the 2004 constructors' championship were allowed to run three cars in the first free practice session. This is why de la Rosa, the reserve driver, was able to have a go.

University in her native Canada, she moved to the University of Cambridge, where she earned her PhD in modelling granular materials. Then, in 1997, she saw an opening at McLaren. 'I thought, "Oh, this could be interesting!" and sent my CV. Someone I'd done my PhD with had been doing some consultancy for them but had moved to the US. They called him up and asked if he knew me, and he said, "Yeah, she can do my job!" They rang me up the same day.'

Hargrove's project was to mathematically model a car: a real-time system that looked at how it moved and acted on track. Then instructions came down from McLaren's managing director, Martin Whitmarsh. Having previously worked in aerospace, Whitmarsh was convinced that there was no reason you couldn't take a flight simulator, such as those used to train pilots, and simulate a racing car instead. 'Adrian Newey had just joined as designer,' Hargrove says. 'He just said, "It's never going to work." Well, that was great for me! When someone of Adrian's stature says it's never going to work, I just say, "Well, I'll give it a go!"'

At the time, McLaren was having its greatest season since the glory days of Senna, winning the constructors' and, with Mika Häkkinen, the drivers' championship. 'Oh, it was unbelievably fun!' Hargrove laughs. 'There couldn't have been a better season for me to go racing!' While Häkkinen won on track, Hargrove worked at the team's factory in Woking, Surrey, trying to perfect her model through programming. Unlike almost everything else in Formula One, the team didn't expect immediate results. 'We knew it wouldn't happen in six months,' Hargrove says. 'We wouldn't have given up, but we knew we needed to be patient, that we'd be going through several different iterations. We didn't spend a huge amount of time on simulators at first because our work could be used by race engineers: it was a car-development tool, not a driver-training tool. A simulator would just be a happy by-product. But it became my passion and not long after I was working on it full time.'

While CFD modelling focuses on air around a car, Hargrove's model was built to look at how the car itself functioned. 'We wrote some of the code from scratch in C++, but the rest we did it in MATLAB and Simulink.' These were packages that had existed since the 1970s and could be used to model a dynamic system – something that changes over time, such as the swinging of a pendulum. 'Some of it was straightforward. It was just modelling things you learn in school, like linkages and springs and so on. And because Formula One cars are so well made, the engineering is so good, you can trust the numbers: your model will resemble reality. Everything was great apart from tyres. Ugh! Modelling tyres is a black art. It's so hard because a tyre's composites, rubber, temperature and even the way the ply is done can change everything … '

By the turn of the millennium, Hargrove's model was so precise it could predict what would happen if the team changed out a single part. Now, the model could be programmed into a basic driving rig, with its steering wheel offering an appropriate level of resistance and the seat rumbling about to mimic how a section of track would feel. This is even more impressive when you keep in mind the technology of the time. There was nothing like this in the world.

'The results were really satisfying,' Hargrove remembers. 'But we were getting complaints from drivers about the graphics. Again and again, it was the graphics! Luckily, computer games were helping us get those better, even if it was pretty difficult at first.' In a fortuitous twist, Hargrove's work had coincided with the release of some of the most realistic race games ever designed, such as *Grand Prix Legends* in 1998 – the forerunner to today's *iRacing*. 'The computer games were helping us get it right. But I remember, many times, David Coulthard saying that his Xbox games were better than the simulator's graphics.* It was frustrating, but we were getting there. We just had to do all kinds of tricks ... '

★ ★ ★

Silverstone is more than a track. As you approach, cutting off the M40 and heading past Brackley towards the circuit, you begin to see signs for just about every fast car manufacturer in the world. This is the northern end of the UK's motorsport valley, with the likes of Red Bull, Williams, Aston Martin, Alpine and Mercedes all within a half-hour drive of the track. An RAF bomber station during the Second World War, it's been used for racing since 1947 when a group of locals broke into the disused field and sped along the runway, accidentally killing a sheep in the process. A year later, the Royal Automobile Club took over and turned it into the home of British motor racing. As the site of the inaugural Formula One race in 1950, you could argue it's the birthplace of the sport worldwide.

Far from being closed to the public, Silverstone is open most of the year. Instead of turning left into the circuit, though, I'm

* This is my second-favourite Coulthard anecdote after one in his book *The Winning Formula* when, during the 1995 British Grand Prix, his radio accidentally picked up a local taxi service's band and he was asked to do a 2.30 pick-up in Towcester. 'I'm sorry, I'm a wee bit busy at the moment ... ' he replied.

turning right, into an unassuming industrial estate populated by rows of large, silver-clad prefab warehouses. The instantly recognisable sign for Lotus sits over one of the units and a small armada of Elise roadsters are parked outside. Evidently, a company car is a perk of the business. My destination is next door: a little more unassuming, with an electric car parked in front of the entrance getting charged. It's only when I step into the reception and see the rows of gleaming silverware that I'm sure I've found my destination.

It's the home of Envision Virgin Racing. Today, they're going to let me play with their simulator.

Downstairs, the team's headquarters is almost space-age in design: shiny white floors, white drawers with tools packed away and an open space that contrasts against the cars' dark blue bodywork. Upstairs are meeting rooms, offices and the simulator itself. It's spread over two rooms. The first is mission control – almost a re-creation of Valencia's telemetry hub, just relocated into a Northamptonshire office – where a gaggle of staff can monitor results. Next door, in dimmed lighting with a wall-sized screen curved around it, is the monocoque of a Formula E car, its bodywork sitting on a raised platform with a series of hydraulics. It's not just a replica of the real thing; it *is* the real thing, just without its wheels. Climb inside and the throttle, brakes and steering wheel all feed your movements to a virtual race projected with utter realism on the screen ahead and around you. It beats anything you can buy at home.

Against the wall, the custom seats of Sam Bird and Robin Frijns wait, ready to cradle their usual occupants. These aren't merely for comfort. Made from a carbon-fibre honeycomb and moulded to the driver's body, each seat is designed to provide a firm, protective shell. The seat itself isn't fixed into the car at all; instead, it's attached at four points, with the driver's weight holding the seat in place. In the event of a crash, the medics can undo the driver's seat belt, then fasten the driver into this ready-made cradle using extra straps tucked away. Once a spine board is inserted to protect the neck, the medics can then lift the driver, still in their seat, out of the car for medical attention without risking further injury.

There's no worry about that today though – any crash will be purely virtual. And there *will* be crashes. The simulator is being driven by a pack of journalists, unaccustomed to the huge acceleration and speeds of a Formula E car. Bird is in the control room, keeping an eye on proceedings and making sure no one gets close to his own lap times. There's bad news for me, though.

'Uhm. We think that you might not … uh, fit.'

'You're worried I'm so fat I'm going to break a titanium beam that can take the weight of every other Formula E car on track?' I ask, pointing to the halo device that fronts the cockpit.

'No, no. It's not that. It's just that you're … uh, big.'

Race drivers are generally small and lithe; I am a 6ft 5in former rugby forward for the University of Bradford. I can't parachute. I can't go in small planes. And now I can't go in a Formula E car.

'It's fine,' I say, trying to hide my crushing embarrassment and disappointment as best I can. 'Let's just talk about how a simulator works instead.'

I can't help but laugh. First, I got my arse wedged while scrambling out of the world's fastest electric car; now I can't even attempt to squeeze my buttocks into a monocoque. It's a fair assessment of my physique, though. After the races in

Valencia, Sam Bird climbed out of his protective jumpsuit, got into a pair of shorts and a T-shirt and ran a half-marathon around the length of the circuit to keep trim. I went back to the hotel and ordered a pizza.

This is not vanity on Bird's part; the rigours of driving simply demand it. 'The forces in a car hit your core, your legs,' Bird says. 'I get a whole-body workout just sitting down and not moving. Which I guess is what everyone wants, isn't it?'

This doesn't delve into the intense environment drivers can face. During a race, a driver can lose 4kg from sweat. 'When I do a four-hour stint in Le Mans in a Ferrari and it's 40°C in the cockpit, that's nasty, that really is,' Bird says. 'When there's no water left in the car, it's tough physically and mentally. I've come in before, curled up on the ground and hugged the concrete to cool down, literally fatigued past my end point. It was at the Nürburgring, I just laid on the shiny concrete floor, curled up in a ball and told my buddy not to come near me for 10 minutes: "Don't touch me, I just feel sick." But it's fine – it's the job. I get paid a salary to be fit and drive that car as fast as I can.'

While the simulator is as close to reality as possible, the staff at Virgin's head office don't subject Bird to temperature-induced water deprivation.* Instead, they focus on familiarisation. 'If a regular person jumped into a Formula E

---

* Sport has tried this before, however. At the marathon for the 1904 St Louis Olympics, the organiser wanted to run a water deprivation experiment on the athletes. He therefore insisted the entire course, run in dust at temperatures in excess of 30°C, only had two water stops. Multiple competitors ended up being taken to hospital, and the eventual winner only got around by being repeatedly injected with strychnine by his trainer. The 1904 Olympic marathon also has an unlikely link to motorsport: the person who crossed the line first, although later disqualified, had hailed a taxi and used it to skip half of the course.

car,' Bird explains, 'it'd feel very alien. You'd feel in too low a position. You wouldn't have any idea what's going on with the screen on the wheel, either. It'd just look like scientific junk, although it makes perfect sense to me. Speed, acceleration and deceleration would be nothing you've ever seen before unless you're a fighter pilot. And with the wheel, I've got hundreds of things I can change at any one time – I can tweak the handling of the car at every stage of a corner using very intricate software and run the track as I see fit.'

Once you're over that initial learning curve, simulators have other roles to play. 'Because track time is limited, simulators are now key to prepare for an event, allowing you to try different things,' explains Ander Fraile González, Virgin's simulation engineer. 'The tricky part isn't trying the set-up, though. It's making these environments as similar as possible to the real thing, not just in terms of graphics but also software complexity.'

This is where the tricks Hargrove mentioned come in. The Formula E monocoque doesn't just move, it reacts to cues in the programme. 'We have butt-kickers that provide vibration if you hit a kerb,' Fraile says, 'and the seat belt is designed to pull when you hit the brake, which fakes the feeling of deceleration.' While this is useful, it's really only the finer details. The real problem is that the image on a computer screen is a 2D projection: you have to trick the driver's brain into seeing three dimensions. To do so, a model has to include everything, from the layout of objects to where the driver 'is' in the virtual world, which means modelling the kinematic chain – basically all the different moving 'bits' of a body – and then doing *inverse* kinematics to make it seem like the body can't just push through the ground, wall or something else. Once you also get the spatial awareness and orientation right, you're beginning to make the simulation feel as real as possible. You then just have to worry about texture mapping, ray tracing, frame rates, sound effects …

'We spent a lot of time working on it,' Hargrove remembers, 'but every time, the drivers would complain. You'd fix the first problem and they'd have another complaint, and you'd work on that instead. But, ultimately, it was satisfying because we were making progress.' The engineers steered the sim; the sim steered the drivers; the drivers steered the engineers. It was a cycle that benefited everyone.

Perhaps more intriguingly, simulations have also opened up an entirely new avenue for people to get into motorsport. Not everyone is lucky enough to get to drive a Formula One car, but most people can play one in a game. And that creates opportunities that previously wouldn't have existed.

★ ★ ★

The cost of motorsport is astronomical. Time in a car, on track, in the right clothes, with the right licences is expensive. It prices most people out of the market instantly. It's why racing dynasties are established, why the 'gentleman driver' exists and why teams often pick drivers not because they're quick but because they can help cover the costs of the sport through sponsorship deals and wealthy benefactors. This doesn't mean that exceptional talent can't fight its way into the sport from modest beginnings. It just means, like Bird, their parents probably had to remortgage their house to get their kid the basic opportunity to drive.

This, of course, creates privilege. You are more likely to reach the top echelons of motorsport if you are rich, and, therefore, far more likely to do it if you're a white male. While nothing can be taken from any driver who does make the pinnacle of elite sport, you're fishing from a small pool. It's entirely possible that the person with the greatest natural driving ability in the world is a girl born in Uganda. It's just almost impossible that such a driver would ever have her talent discovered. Sexism, racism, nationalism, classism and financial disadvantages act as gatekeepers to motorsport.

But anyone can pick up a PlayStation controller and drive. Today, esports is a huge business and, while staying competitive requires increasingly expensive kit, the cost of a top-of-the-range home system is a fraction of a day on track. Which begs an obvious question: would someone who excels at racing on their home-gaming system also be fast in the real world?

In 2020, 22-year-old James Baldwin showed that they might be. At 16, he was forced to drop out of kart racing because of lack of funds and gave up on his dream of being a pro racer entirely. A year later, walking around PC World, he stumbled on a basic home simulator for £250 and, on a whim, bought it. After a year of playing racing games on his home computer, he entered the world of esports before joining World's Fastest Gamer – a two-week reality-show competition that gave esports players the chance to drive a real racing car. After being picked as a winner by the judges (including none other than Juan Pablo Montoya), Baldwin was given a seat on Jenson Button's GT3 race team, Jenson Team Rocket RJN. In August 2020, he made his debut at the British GT, lining up against seasoned veterans and those with the financial backing Baldwin had lacked.

Baldwin won.

Of course, translating esports to real racing isn't as simple as that – just ask the man who led to this book being written (see the Appendix for details), race driver and team owner Martin Short. After starting late for a racer at 27, he quickly rose through the ranks, winning the British GT and racing at the 24 Hours of Le Mans. When he noticed his 16-year-old son, Morgan, was a keen sim racer, he thought it might be fun to put him on track at Mallory Park and give him a go in a real car. 'First, I showed him what it could actually do speed-wise. That was a really stupid mistake. Then I stopped and asked him if he'd like to do a few slow laps by himself, just to get a feel for the car. The next thing I saw was him going around the Gerard's Bend – a huge, long, massive radius exit – at something close to 100 miles per hour.'

On the third lap round, telemetry shows Morgan took Gerard's at the exact same line as his dad, at exactly the same speed. Unfortunately, his wheel touched the oily part of the track, just off the racing line, and he began to lose grip. 'And that's what you don't get in the simulator,' Martin explains. 'You don't get the feeling in your arse that tells you that something is going wrong. The rear of the car started to slide left and started to point toward the infield of the track. And then Morgan headed infield. He didn't even put his foot on the brakes. He didn't know what to do in that situation, despite doing it ten thousand times on a simulator, and he panicked. His lack of knowledge and experience meant he was completely unprepared.'

This is where sims collide with real life – and where Morgan collided with a dirt barrier at 85mph, landing on the rear suspension so hard it snapped in half. 'It was awful,' Short recalls. 'I was running over there, worried for my son. I'd thought everything I'd gained from thirty-three years' experience would be completely obvious and natural to him, because I'd seen him look so good in a sim. I should have done so, so much better. I'd feared, as a dad, being overbearing and it all went wrong in the most horrible fashion. It could have been so much more serious.'

Morgan was fine. Today, he's one of the top esports racers in the world – going toe-to-toe with Baldwin. He's also a promising young racer in real life; after proper training, he was able to beat his pro-racer trainer, Ian Flux, around Silverstone by three seconds. But the story illustrates that racing in a sim, no matter how real it seems, isn't quite the same as reality.* It reminds me of the movie *Aliens* when hero Ripley asks the nervy Lieutenant Gorman about his experience. 'How many drops is this for you, Lieutenant?' Gorman replies he's been on 38 … simulated. 'How many *combat* drops?' asks one of his Marines. 'Uh, two. Including this one.' It's not a huge spoiler to say Gorman's inexperience doesn't bode well for the team.

It's also worth remembering that real life isn't the same as a simulation when it comes to data, too. In 2020, Formula One and Amazon's Machine Learning Solutions Lab spent a year building an algorithm that could work out the fastest F1 driver of all time. And, while the top three were fairly predictable (Ayrton Senna, Michael Schumacher and Sir Lewis Hamilton), I doubt many were expecting Heikki Kovalainen and Jarno Trulli, both of whom only won a single race in their F1 career, to come in at eighth and ninth in the all-time rankings.

---

* Race drivers can get in trouble in the virtual world. In May 2020, IndyCar's Simon Pagenaud was lambasted for deliberately crashing into McLaren's Lando Norris during a virtual race. And, later the same month, Audi Sport's Formula E team sacked driver Daniel Abt after it emerged he'd asked pro gamer Lorenz Hoerzing to take his place in a virtual charity event. Sim racing is no longer a game.

Despite these limitations, though, the lines between racing in the real and virtual worlds are blurring. And this is only a fraction of what Formula One simulation tech has led us towards. Believe it or not, there's a direct link between David Coulthard's Xbox gripes and some of the most exciting innovations in modern healthcare.

* * *

In the early 2000s, it became obvious the McLaren simulator was doing its job. Hargrove's team began to expand. So, too, could the simulations, aided by driver feedback and real race data. 'We could think about making a change to our suspension this way, trying different lines, learning and adapting our designs,' she says. 'We were the only ones with a simulator, so it was a big advantage for a number of years.'

Eventually, one simulator became two, which was rented to other Formula One teams until they also adopted the technology. Today, every team has its own simulator, both for research and driver training. Hargrove, meanwhile, joined the newly formed McLaren Applied Technologies, first as programme director and eventually chief technology officer. She then left racing to become chief technical officer of Babylon Health, migrating the skills she honed creating simulators into a whole new area: making a virtual human.

Hargrove explains the concept of a digital twin using the *Apollo 13* disaster from 1970. Stranded near the Moon, the astronauts needed a solution to fix damage from an exploding oxygen tank. Back on Earth, NASA created a mocked-up version of the *Apollo 13* modules, workshopping different options with what the astronauts had available. The digital twin is exactly the same, only for your body: a computer version of you, which can be poked, prodded and monkeyed with to try and keep you healthy.

Simple versions of this concept have already been used in motorsport. Since 2008, the FIA Institute, the FIA's safety arm, has been working with Toyota to improve safety in high-speed collisions. This is done through a simulation called the Total Human Model for Safety (THUMS). Essentially a virtual crash test dummy, THUMS models a human body in a road traffic accident, allowing researchers to predict what might happen to a person's bones, muscles, ligaments, tendons and internal organs. As with Hargrove's simulations, THUMS has grown exponentially as processing power has expanded. When it was first created in 1997, it consisted of 80,000 digital elements, taking a whole day of processing time to run a simulation. By 2012, the model had 2.2 million elements, with a precise model of the brain based on CT scans. This detail has informed car design should a crash occur. For example, the FIA found that changing a seat's angle from upright to a position between 40–60° reduced forces on the spine during an impact by 22 per cent – which is why today you'll see racing car seats at an angle. THUMS is currently used by Formula One, IndyCar and NASCAR to attempt to minimise injuries, along with more than a hundred vehicle manufacturers, suppliers and research bodies to save lives on the road. In 2021, the software was released online for free, meaning it's likely to become a standard tool for decades to come.

While THUMS is impressive, Hargrove's dreams go even further. Rather than use a generic model of a human, her aim is, eventually, to create an exact digital double of an individual – a virtual version of *you*. Already, data from biobanks and anonymised medical reports has allowed health scientists to understand who we are at a genetic level. But such information lacks context. A digital twin, on the other hand, would enable doctors and scientists to see how your body functions, down to each individual cell. Your twin could then be put in a range of different situations and environments to see how it would affect you.

This could be general or specific. General because, in a world where more people have access to a mobile phone than to a doctor, you can give this information to everyone and make healthcare more equitable across the globe. 'In the UK or Canada, we take our healthcare systems for granted,' Hargrove explains. 'But that's not the norm for someone in the US, where good hospitals will do too many tests to cover themselves, often at a patient's expense; or in somewhere like Africa or Southeast Asia, where there's a cultural belief that the most expensive hospital is the best, even though it might not be.' Put an app in the palm of a patient's hand that shows them how their body changes and they can make more informed choices about their health without worrying about the bill.*

Specific, because how your body works is unique. Perhaps you work up a sweat easily; perhaps past military trauma means you can't sweat at all. A virtual twin can take such details into account, allowing medics to tailor drug doses and timings to make sure you're getting the optimum benefit with the minimum side effects. With a digital twin, you could even predict how your body will be affected by ageing, disease,

* Sadly, there isn't enough space here to go into the huge ethical, political, financial and philosophical ramifications of this technology, which is a book unto itself. Hargrove is aware of them.

exercise and diet, allowing potential risks such as cancer to be identified decades in advance.

'We've only done a little of this so far, as there's a lot of groundwork to cover,' Hargrove says. 'We're taking baby steps. But the ambition is there. Right now, we are focused on preventative medicine, trying to give you your risk in five years' time of developing diseases that you can change through your lifestyle. Eventually, we want to do this much more proactively. For example, we've developed a few simulations for people trying to lose weight – what will make a difference for *you* and *your lifestyle*, rather than just the general guidelines people are given at the moment. We can also do ongoing monitoring. What if we can help you sleep better, so you might not feel so tired and therefore you've got more motivation? We aren't looking at crude things, just one lever of health. We're trying to bring all of healthcare together. At times, I'm completely discouraged. I think "How are we ever going to crack this?" But actually, that's why we need to spend time on it.'

Twenty years ago, Adrian Newey bet against Hargrove. I'm not going to make the same mistake. Since leaving Babylon Health in 2021, she has joined start-up Zedsen to develop sensors for projects such as non-invasive blood glucose testing or identifying cancer tumours. It's yet more valuable data, she says, for building better digital twins in the future.

It's astonishing to think, standing in the Envision Virgin Racing offices, where a Formula One simulator has led us. Another time from a journalist flashes up in the control room and one of the engineers looks at Bird. 'He's quick,' she says. 'A few more laps and he might be close to you.'

There's a knowing laugh. 'I'll just have to climb in and have a go myself,' Bird replies. I'm not sure whether he means it, but either way I can't help admire his dedication to perfecting the racing art. For him, as with any other elite athlete, this isn't a job for a few weekends of the year. It's something that flows deep in the blood. Racing is something you eat, sleep and breathe.

'Don't you ever get bored, going around the same track all day?' I ask Bird.

He pauses and briefly frowns for a few seconds at what might be the dumbest question anyone's ever asked him. 'No,' he says, shaking his head, unable to even imagine it. He squeezes past, out of the control booth. 'I'm not going to lie, I love my job.'

He wasn't kidding. A week later, he won the Formula E season opener at Diriyah in Saudi Arabia. It was only the second time the veteran driver had raced on it, but probably the thousandth time simulated. I had planned to catch up with Bird and the Virgin team again, a few races later in Rome, but it never happened.

The Covid-19 pandemic struck. The season was suspended.

And, as most of us sat at home facing an uncertain future, the greatest race in motorsport history began.

PART TWO

RACING FOR LIFE

CHAPTER EIGHT

The Full Might of What We Can Do

Professor Becky Shipley walked into the soft, comfortable interior of the staff common room at University College London (UCL) and sank into one of its deep crimson leather sofas. It was Tuesday evening, 17 March 2020. Outside, the leafy squares of Bloomsbury were placid in the evening air, the streets empty and eerily quiet even for the renowned peace of this oasis in the heart of London. The Covid-19 virus had spread across the globe, migrating in bodies of travellers, unknowing vectors who took a modern plague across borders and continents. The virus made you struggle for breath; the worst hit couldn't breathe at all.

Gradually, the world was waking to a bitter truth. It faced a global pandemic in which millions would die.

A few days earlier, the UK government had launched its 'Ventilator Challenge'. The UK's hospitals had fewer than 6,000 ventilators available at the start of the pandemic and, after a rush to get more, it still only had 8,175; to meet the expected worst-case scenario for Covid-19, it needed 30,000 in a matter of weeks. The scramble had been unprecedented; the National Health Service (NHS) had even taken functional ventilators used as props on the TV medical drama *Holby City*. The Ventilator Challenge was recruiting groups across the UK's manufacturing sector, asking them to stop business as usual and start saving lives. That Sunday, Shipley had been called by Professor Cathy Holloway, who had been asked to lead one of the government's low-cost ventilator 'design sprints'. Would Shipley like to run the project with her?

Shipley had declined.

A bioengineer, Shipley had been raised in nearby Buckinghamshire before studying mathematics at the University of Oxford. Taking an interest in the human body, she had completed her doctorate in modelling drug transport, developing advanced tissue simulations focused on blood flow in cancers. She had also taken an interest in repairing the nervous system, using computational design to understand how our bodies respond to stimuli. Now at UCL, she was director of the university's Institute of Healthcare Engineering. And she knew the government's plan wouldn't work.

Instead, she had called an informal meeting in the common room. Joining her on the sofa was Mervyn Singer, professor of intensive care medicine at University College Hospital (UCLH) across the road. Sitting opposite was Tim Baker, a UCL professor of mechanical engineering. Rather than come up with new ventilator designs, the trio realised UK hospitals would be overwhelmed in weeks unless they could keep people *off* ventilators. Singer even had an idea for a machine that could do it. The group just needed advice on how to make the thousands of units necessary before the pandemic hit its peak.

Now Baker took the lead. Before joining UCL, he had been part of several Formula One teams, as well as teams that won their class at 24 Hours of Le Mans and the World Rallycross Championship. Picking up the phone, he rang an old friend and colleague, Ben Hodgkinson, the head of mechanical engineering at Mercedes AMG High Performance Powertrains, responsible for the power units for the Mercedes, Williams and Racing Point teams. Baker told Hodgkinson their idea and asked if he could take a sabbatical for a few days, come down to London and consult on their project. Motorsport engineers, after all, were used to pulling off the impossible.

Hodgkinson took the plan to his boss, managing director Andy Cowell. His response to the UCL professors was even more ambitious: 'Do not hesitate to call upon the full might of what we can do.'

UCL and Formula One's leading engine manufacturer were about to pull off one of the most astonishing, life-saving feats in engineering history.

★ ★ ★

Formula One has a surprising connection with the NHS. While outsiders might think it's largely one way – with doctors eagerly signing up to work as part of the track safety teams in races – the reality is far different. For the past 20 years, Formula One has been lending its expertise to doctors in all sorts of unlikely ways.

In 2003, paediatricians Allan Goldman and Martin Elliott were watching a grand prix after their shift at Great Ormond Street Hospital. Both realised there was a startling similarity between a pit stop and transferring a child after a successful operation to the intensive care unit. In hospitals, 70 per cent of mistakes are usually breakdowns in communication – technical or information errors – and half of these occur at this 'hand-off', for example by transferring a patient before a bed is set up for them. Goldman wondered if his team could copy the balletic, choreographed precision that makes a pit crew able to change four tyres in less than three seconds.

(The world record, in case you're wondering, is held by Red Bull Racing: a blistering 1.82 seconds at the 2019 Brazilian Grand Prix.)

Goldman asked McLaren to explain how it choreographed its pit moves, then headed to Maranello to witness the Ferrari mechanics in action. The lessons were obvious: strict, well-drilled routines practised to perfection and reviewed; each person with one task, working in calm silence rather than distracting noise; and a 'lollipop man' to coordinate the effort. The team sent Ferrari a tape of their own handover; the mechanics took notes, made suggestions, sent them back. The Great Ormond Street surgical teams listened, working with Ferrari and airline pilots to come up with new handover protocols. This included assigning one member of staff to coordinate and sign off the effort – their own version of the lollipop man. After two years, the surgeons published their results: technical errors had fallen by 42 per cent, information errors by almost half. Operation Pit Stop had been a complete success.

The collaborations haven't ended there. In 2015, McLaren Applied Technologies partnered with the University of Oxford to improve surgical training, clinical care and hospital facilities. One of the first steps was to introduce data to predict the best trainees. For centuries, prospective surgeons had been assessed by eye alone, but McLaren worked with Oxford to create a sensor, attached to a surgeon's elbow, which tracked hand and wrist motions. 'There's a spot on your arm where, if you put a very good accelerometer there, you can see what finger is moving and how it's moving,' Caroline Hargrove says. 'By putting sensors at the top of the forearm of surgeons, trainees as well as really good ones, you could see what differentiated them in a number of different tasks and metrics. Which is pretty cool.' Now trainees had real-time feedback on their actions and examiners could use an objective, evidence-based number to compare dexterity and skill. In the early 2000s, baseball scouting had been transformed when 'sabermetrics' were introduced. This used maths and percentages to choose which players to sign, pitches to throw and batting swings to adopt to bring the team success. Sabermetrics was a game changer: so much so it was turned into a Brad Pitt film, *Moneyball*. Thanks to Formula One, sabermetrics had just entered modern surgery.*

Hargrove remembers using remote sensors for drug trials, too. 'We were approached by Pfizer, who asked if we could monitor patients remotely, in the same way we monitor a car during a race.' Rather than visiting patients at intervals, as is typical in a clinical trial, the company wanted to get a bigger, cheaper picture of their patients' health as they developed new medicines. This didn't just save money and identify problems with a drug sooner: by constant monitoring, the trial would also be able to pivot and make adjustments that could produce better results. 'It was the first time I worked in health,' Hargrove recalls. 'We sourced all these sensors from wellness and sport, and made them transmit wirelessly to a database where the data could be analysed. We then did it with another pharma company, which was the first clinical trial that had remote monitoring with it.'

---

* As anyone who works in data knows, this kind of work can introduce unconscious biases: if most 'good' surgeons are old white dudes, are you unfairly excluding students from other demographics who might move and act differently? This was something McLaren actually considered during their data analyses to ensure that the results were objective and not discriminatory.

McLaren was also at work in the wards of Birmingham Children's Hospital. In 2008, Consultant Heather Duncan had created a system of predicting risk signs of a heart attack for children in hospital – a problem affecting 650 kids in the UK every year. While life-saving, Duncan's system was far from ideal. The child had to be hooked up to wires and sensors, so they couldn't leave their room, and recording the information had to be carried out manually, with nurses coming by to jot down vital signs every four hours. Duncan needed a remote system that could monitor her patients in real time. And, while such devices existed for adults, there wasn't an equivalent for children, whose growing bodies and different physiology meant that no child was alike. Any remote monitor needed to be able to learn what was 'normal' for an individual child and pick up on subtle changes that could spell danger.

That's when Duncan crossed paths with Formula One. 'It was completely by chance,' she told the BBC in 2012. 'I was at a conference about detecting illness in hospitalised patients. Peter van Manen, the MD of McLaren Electronics, was doing a similar presentation about what we could learn from the health of Formula One cars. It became clear they'd already solved a lot of the issues that were so challenging for us in trying to specifically monitor children.' With McLaren's help, Duncan and colleagues were able to create a system that not only set off an alert when a child was at risk but also learned and adapted to the individual patient's needs. It was an incredible advance, although it still had a few quirks. 'You can tell this is Formula One technology,' Duncan said, while showcasing the vital signs of a little boy playing in the corner. 'Here we can see Damien is on his "out lap", and, if we really wanted, we could try to measure his brake temperatures, tyre pressures and how much fuel he has left … '

By 2015, the next stage of the device was already in use: the Real-time Adaptive and Predictive Indicator of Deterioration (RAPID). McLaren had developed a system that could transmit children's vital signs wirelessly via Bluetooth.* Now the children weren't anchored to their bed with multiple wires on their body – they could explore and play. After three years, more than 1,000 children had used the device, starting with four-year-old Maci Walford, a little girl with a congenital heart condition. 'This technology is truly transformational,' Duncan said. 'I genuinely believe this will change the way we care for patients in hospital forever.'

Given F1's pedigree in solving problems, it's hardly surprising Baker knew the Mercedes team had the ability to help with the UCL plan – and it wasn't like they were busy. The Formula One season was on an unofficial hiatus. That weekend, the Mercedes team had been in Melbourne, Australia, preparing to defend both the constructors' and drivers' championship titles, with reigning champion Lewis Hamilton once again in the lead car. But three days before the season's opening race, three members of the McLaren team had tested positive for Covid-19. The race was abandoned, joining the already cancelled Chinese Grand Prix. Further delays and cancellations had begun to trickle through, with some races barring spectators and others shifting further down the calendar. By the time Baker phoned Hodgkinson, it was apparent that there

* McLaren wasn't solely responsible: it was working with Birmingham Children's Hospital, the University of Birmingham, Aston University and Isansys Lifecare.

would be no racing until May at the earliest. The engines were off, the garages shut, the pit wall empty.

Even so, Cowell's unflinching response to the UCL professors was an astonishing commitment. This wasn't about applying existing Formula One principles to another environment, or adapting existing technology for a greater purpose. He had just put the full strength of a Formula One team into making a healthcare device they'd never seen before, let alone used in racing. Heading down to London, Hodgkinson met the UCL and UCLH team, which handed him the device they had in mind. Immediately, Hodgkinson recognised the elegance of its simple design. It was valves. Hydraulics. Simple mechanics.

It was just the kind of challenge his engineers would relish. He picked his best three and told them they had two hours to get to London.

★ ★ ★

A ventilator is basically a high-tech set of bellows that automatically responds to a patient's breathing, taking over and doing the hard work for them. It pumps air (or oxygen) into your lungs through a tube that passes via your mouth into the wind pipe (called an endotracheal tube) and gets rid of waste gases such as carbon dioxide. It's great at helping a patient breathe and is quite simply the best hope for keeping a very sick Covid-19 patient alive. The downside is that ventilators are big, complex and require specialist skills to design and make – particularly when it comes to their control mechanisms. 'The idea that you could redesign and manufacture them at scale in weeks didn't make much sense,' Shipley recalls.

It was just one of three major problems the UCL professors had spotted with the government's Ventilator Challenge.

The second issue was that, even if the country pulled off a small miracle and managed to get the machines made, there weren't enough people trained in how to use them. 'It's all very well having thirty thousand ventilators but you need the staff to look after them – which we don't have,' Singer told students at St Edward's School, Oxford as part of their *The Ventura Project* podcast series.* 'In the UK, we have four thousand intensive care beds and would need one nurse per patient. All of a sudden, you'd have lots of machines and no one to run them.'

Even if enough nurses could be found (and trained) to use the machines, there was a final problem: ventilators are *horrible*. As they involve slipping a tube down someone's throat, you have to keep the patient sedated. That means they need round-the-clock care, IV bags and catheters. And, as with any invasive procedure, there's always a significant inherent risk. For ventilators, risks include infection, scarring the lungs and even psychological trauma from the deeply unpleasant experience. While patients going on ventilators are, of course, very sick, the figures are stark: 50 per cent of Covid-19 patients on a ventilator die; 40 per cent need dialysis and 30 per cent cardiovascular support. You do not want to end up on such a machine.

'Mervyn and Dave [Brealey, another intensive care consultant at UCLH] have this amazing network of critical care colleagues around the world and had been talking to

---

* The St Edward's students' four-part podcast of the entire story is awesome. You can check it out by following Celiafilms on Soundcloud.

people in China and Italy,' Shipley continues. 'Their experience was they had been overwhelmed very quickly, because once a patient's on a ventilator they are on it for weeks. They need intensive nurse support, renal support and pulmonary support. Hospitals just run out of space very quickly.'

Singer's alternative was to use a continuous positive airway pressure, or CPAP, machine. Typically used to treat acute respiratory diseases, or even snoring, a CPAP machine is simple to operate and patients only need to wear a tight mask for it to work – no nasty tube down their throat, no extra risks, no special training of staff required. Rather than function like an artificial lung, a CPAP uses the Venturi effect (fast-moving air sucks the air around it) to push gases through a series of valves and create a rapid air flow. 'Basically, a CPAP takes pressurised air from the hospital's oxygen supply,' Shipley explains, 'and inflates your lungs a little bit, allowing you to get more oxygen.'

Think about blowing up a balloon. At first, it takes a lot of effort. But, once you've got a little air inside, the balloon inflates far more easily. The added pressure from a CPAP machine is like the first puff to get the balloon going. 'The alveoli of the lungs are like loads of little balloons,' Shipley explains. 'The CPAP is holding them open and that means more oxygen can get in them, which means you can get more oxygen into your bloodstream.'

While CPAP machines couldn't replace a ventilator for the most critical cases, they had already been used in China and Italy to help less serious patients who still needed oxygen. And they *worked*. Singer's Italian colleagues reported that using CPAP machines kept around half of their patients out of intensive care and off ventilators. Better still, their

CPAP patients usually spent eight days on the device, rather than up to a month on a ventilator. Not only did a CPAP machine improve a patient's chance of survival, it meant more beds and resources would be free if the Covid-19 crisis deepened.

Unfortunately, there were only a handful of CPAP devices in the UK. UCLH Trust had just 12. The team knew the UK's hospitals would need thousands of machines if they were going to make a difference. And they needed them urgently.*

'We didn't have long,' Shipley recalls. 'This was on 17 March. The peak in London was expected on the Easter weekend [12 April].' On such a timescale, it was impossible to design a new CPAP machine from scratch. Instead, Singer suggested an alternative: take an existing model, previously used in the NHS, that could be dusted off and recommissioned. 'We decided to go back and use the Philips Respironics WhisperFlow, which is an off-patent, CE-marked [made to EU safety standards] device,' Shipley says. 'There was a good evidence base, both in the UK and internationally, and it's really simple to use: there's no embedded electronics, it's purely mechanical. We thought if we went back and reverse-engineered one, we could make them quickly enough. We wouldn't be going from a standing start.'

The team tried to get hold of the plans for the device, but failed. 'We tried to reach Philips but we didn't have the

---

* This is not a failure of the NHS, which is, as far as I'm concerned, one of the wonders of the modern world. Remember: before Covid-19, there were more than enough ventilators and no reason to have a load of CPAP machines lying around.

contacts,' Shipley recalls. 'So, instead, we found an old WhisperFlow in the anaesthesia museum at UCLH.' Once again, it was Brealey who had assisted the team (it turns out he isn't just good at saving lives, he's also great at scavenger hunts). This was the device given to Hodgkinson.

Even with an existing device earmarked for use, the task facing UCL – and Mercedes – was incredible. They had to take the WhisperFlow, reverse engineer and copy it perfectly, design a new version, prove it was safe, get it signed off by regulators and go into full-scale production. It was a task that usually took a couple of years. The UCL team had to do it in less than four weeks. 'We were a bit bonkers, really,' Shipley concedes.

In London, the team based itself at UCL's MechSpace,* with doctors, academics and engineers working together, and the Mercedes and UCL engineers renting nearby hotel rooms so they could toil into the night. Among them were three former UCL students – Ismail Ahmed, Alex Blakesley and Jamie Robinson, the last of whom brought along computer design software he'd created to run the prototype process. Baker and Shipley used their contacts to obtain parts that couldn't be created from scratch, such as sensors and the tight-fit masks. Oxford Optronix, a company that designs parts to measure oxygen flow, was recruited to help as well. The company's founder had been walking his dog when the call came through; in an instant, the walk was over and he was rushing to get his staff mobilised.

---

* This is a very cool name for a base of operations and the reality is even more badass: six storeys of funky white study rooms, workshops, 3D-printing labs and VR suites. It is basically an engineering paradise in the heart of London.

Soon, the team had two WhisperFlows: the one found in the museum and a second Cowell bought off eBay; rather than wait for it in the post, Mercedes' operations manager was dispatched to pick it up from the seller's doorstep and bring it back to the team factory in Brixworth, just outside Northampton, for analysis. In both London and Brixworth, the engineers began to reverse engineer the devices the old-fashioned way: by slicing them in half and poking about. From there, things became a little more high tech. 'Once we worked out all of the design, we redid it in 2D and 3D and ran materials characterisations,' Shipley says. 'We did CT scans, then we manufactured a prototype.'

The group worked round the clock without a break, drafting drawings and coming up with new iterations. Hodgkinson had told his team to come to London without stopping at home, so the Mercedes engineers didn't even have a change of clothes. The UCL engineers were in a similar situation: they had all come into work on Wednesday and suddenly found they'd have to stay put. 'When it became clear it would be more than a couple of nights we'd need to stay, one of the university students was sent out to buy pants, toothbrushes and T-shirts,' Hodgkinson told the St Edward's students. 'The only shirts he could find were some pink ones with "Ventura" written on them.' The emergency apparel became the official team uniform, and the collaboration had a name: UCL-Ventura.

The work was intense. 'I famously said, "This should take us four hours,"' Hodgkinson remembered. 'It took us more like forty. We worked until five a.m. and then got up at seven a.m. for three days in a row. I can honestly say, even in Formula One, I've never worked that hard ... in Formula One I manage my stress by telling myself, "It's only racing cars." When you're working on something that could potentially save lives, you don't have that. I've never pushed myself so hard. We slept for seven hours in three days.'

Hodgkinson had set a marker to have a prototype under way within 24 hours of arriving at the MechSpace. In the end, the team missed the deadline by two hours. Three days later, the first devices were at the hospital, ready for testing by volunteers.

'We started on Tuesday, Mercedes turned up on Wednesday and by Sunday we were at UCLH testing them on the wards,' Shipley recalls. It was one of the fastest turnarounds in medical history, made all the more impressive because it couldn't use modern techniques such as 3D printing in case loose flakes fell into the air flow – the team had to make the device just as it had been designed in the 1960s.

By now, the UK's medical watchdog, the Medicines and Healthcare Products Regulatory Agency (MHRA), was aware of the project. 'We reached out to them that Wednesday [18 March],' Shipley says. 'I remember emailing their chief executive, June Raine, at about ten o'clock at night – we'd been working for a whole day then, and it had become clear we could do this. She replied in ten minutes and we spoke to the MHRA the next day. From that day forth, we were in contact every day, so we had a real understanding of what we needed to do to proceed with approval.'

This wasn't mere rubber-stamping. The MHRA needed evidence – and lots of it. Bench testing to show the new device was an exact replica of the WhisperFlow; trials with healthy volunteers to show it was safe and worked; a roadmap for clinical evaluations; evidence of quality management in manufacturing; a plan for monitoring systems and gathering adverse event reports. The UCL-Ventura team had to produce dozens of pages of evidence, documenting their work and meeting every exacting technical specification. It would be easy to paint the MHRA as a box-ticking villain here but the truth is the opposite: it was always an ally. In his former life as a hospital doctor, medical devices director Dr Neil McGuire had used WhisperFlow devices to treat patients. He saw the potential of what the team were doing immediately. Instead of standing back, McGuire worked closely with UCL-Ventura to address any concerns about safety and advise the team when they were straying from requirements. At one point, that involved telling Hodgkinson to put back rough, ungainly edges he'd filed off the machine – his new, nicer version didn't comply with old British safety standards.

Somehow, in a near-impossible timeframe, the UCL-Ventura group managed to replicate the device perfectly. 'We then had to gather all this data and evidence and submit it to the MHRA,' Shipley says with a slight sigh, recalling the heavy amount of work involved. 'They approved it ten days after our first meeting.'

The WhisperFlow was back. But it still wasn't good enough to help someone with Covid-19. For starters, the device had never been designed to treat infectious disease, so didn't have safeguards to stop patients from exhaling droplets and spreading the virus. This meant that the breathing circuits, which connect the CPAP to the patient via the mask, needed to be redesigned to incorporate filters. There was also a problem with how oxygen-thirsty the new machines were. 'In hospitals, networks of pipes bring oxygen to individual wards and patient bays,' says Shipley. 'And normally there's no way you'd ever reach capacity on that system. But the oxygen demands of treating Covid-19 made us concerned.'

Even as the MHRA reviewed their take on the WhisperFlow device, UCL-Ventura were already working on a second version. Their prototype had been a carbon copy of the original but had little design flaws that irritated the Mercedes perfectionists, with Cowell describing it as a 'well-loved toy rather than an engineering device'. The engineer couldn't resist the chance to tinker and make an even better CPAP machine. He began to suggest improvements they could make to the machine's assembly, including adding a single flap valve to cut down flow. 'Engineers are creative,' Cowell said on the St Edward's podcast. 'We all want to be the *most* creative. That mix of blunt, logical engineering calculation [combined] with a big ego – we've all got big egos, some of us know, some of us are in denial – means we want to have the best ideas. On a simple task like copying, the creative genius gets frustrated and bubbles up in the odd area. Engineers will never just copy, they will always improve … but you have to test to make sure that change hasn't brought with it some unwanted madness.'

'We had to focus on the redesign of the breathing circuits as well as the CPAP machine – tubing, valves, filters, mask,' Shipley says. 'This was a big deal, and was something that came to our attention as soon as we started testing in the hospital. The breathing circuits define the resistance to flow, so make a big difference when it comes to oxygen utilisation. By designing our own we could make them as efficient as possible.'

Above: Camille Jenatzy in *La Jamais Contente* after breaking the land-speed record, 29 April 1899. It was the first purpose-built race car, powered entirely by electricity.

Below: *La Jamais Contente* reborn: the Buckeye Bullet 3 in all its glory, preparing to brave the Bonneville Salt Flats.

Above: The author with the Buckeye Bullet 3 at The Ohio State University. The chassis holds the current electric land-speed record of 341.264mph.

Below: Jules Goux won the Indianapolis 500 in 1913. His Peugeot used four valves in its engine cylinders, allowing him to win the race by more than 13 minutes.

Above: Formula E in action, driven by Virgin's Nick Cassidy at the 2021 New York City E-Prix. These electric cars reach speeds of more than 174mph, helping to develop technologies we'll soon see on our roads.

Right: The author at Circuit Ricardo Tormo for Formula E pre-season testing, October 2019. None of us quite realised what a season it would be.

Right: The author inside a Shelby Cobra at Oak Ridge National Laboratory. The entire Cobra, except the engine and tyres, was 3D-printed from scratch.

Left: Extreme E's Odyssey 21 before team colours are added. The engine is electric, the bodywork is made of Bcomp's flax fibre panels, and the Continental tyres are made from dandelion rubber.

Below: An inside look at the Forze VIII from the Delft University of Technology, Netherlands. The hydrogen-powered team became the first to beat petrol-driven rivals, and are aiming to enter the 24 Hours of Le Mans.

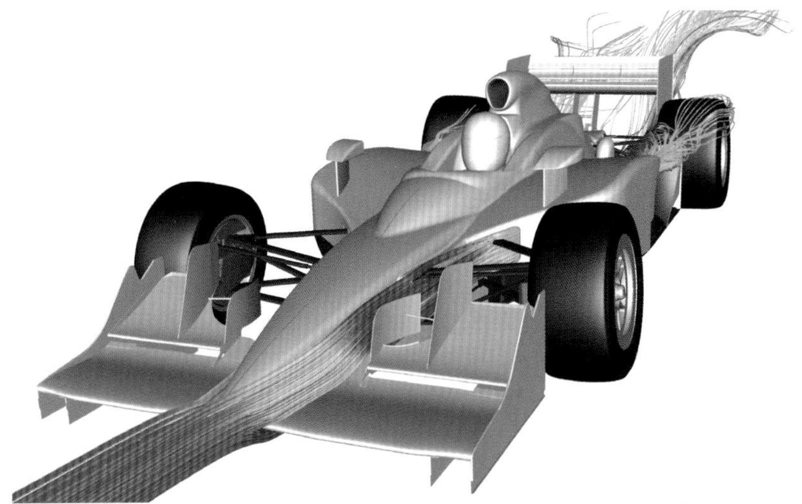

Above: Computational Fluid Dynamics (CFD) applied to an IndyCar. Using computer modelling and calculations, teams can adjust aerodynamics and predict whether air flow is smooth or turbulent. Useful for going faster.

Right: Insights from CFD allow design changes that keep cold air trapped inside supermarket fridges, thus lowering energy bills.

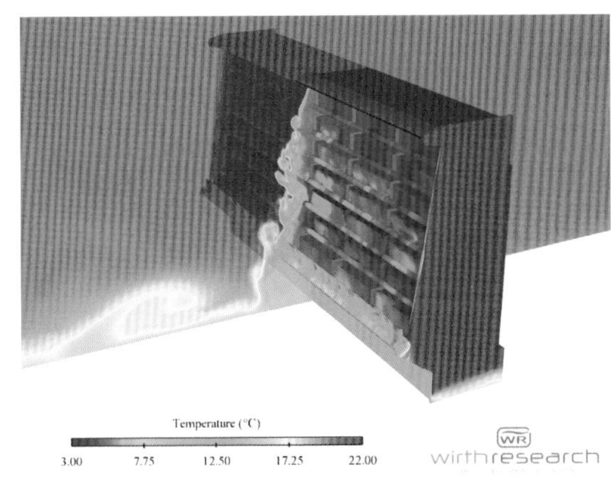

Below: CFD can even model entire cityscapes, such as Chicago, to predict how skyscrapers will affect the wind.

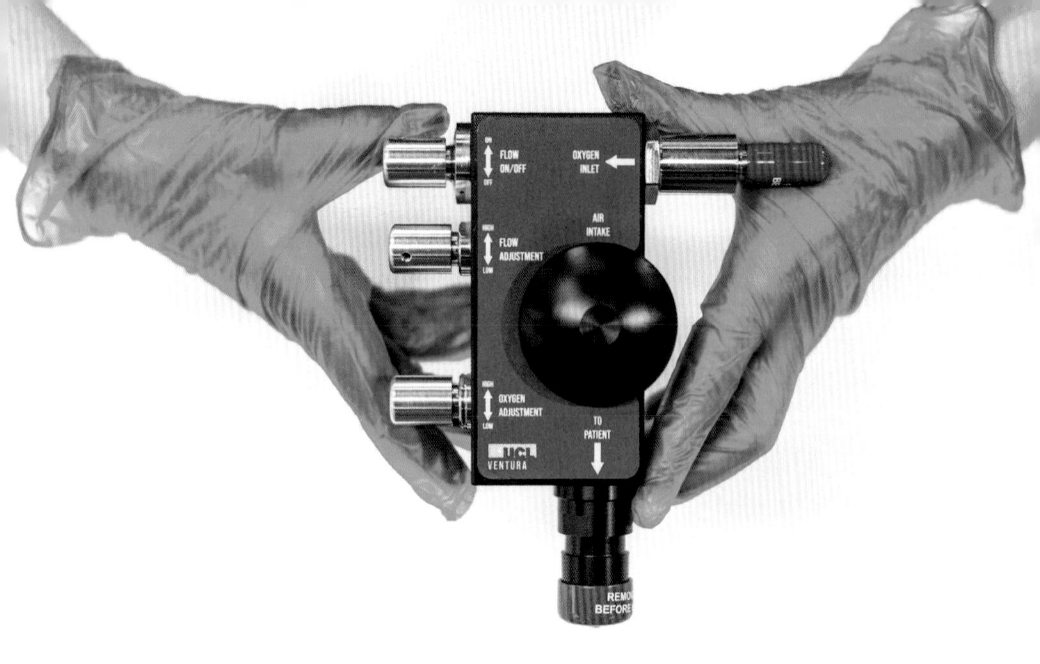

Above: The UCL–Ventura CPAP device. Created by a team led by University College London and Mercedes AMG High Performance Powertrains, the group went from the drawing board to delivering 10,000 breathing devices to UK hospitals in less than a month.

Left: The famous UCL–Ventura pink shirts. The name comes from the fact that the Mercedes engineers (including Ben Hodgkinson, second from right) had arrived in London ready to work for days without a change of clothes ... the shirts were all they could find nearby!

Left: Professors Rebecca Shipley and Tim Baker from University College London discussing the design of the new CPAP device during the COVID-19 pandemic.

Above: Dorothy Levitt, the 'Fastest Girl on Earth'. Her 1909 book *The Woman and the Car* was one of the first to advocate rear-view mirrors.

Above: Hugh Cairns. His pioneering work on crash helmets following the death of Lawrence of Arabia continues to save lives.

Below: Ayrton Senna at the 1988 Australian Grand Prix. Senna's death in 1994 was a pivotal moment in motorsport safety and its subsequent impact on driverless cars.

Above: Robocar. With a two-way speed of 175.49mph, it is the fastest autonomous car in history and a demonstration of what's to come in the future.

Above: Roborace's DevBot 2.0, which has pushed driverless cars to the point where wheel-to-wheel racing is possible and an AI can almost outpace a world-champion driver.

To test the circuits, a new member of the team was brought in: a full-size male mannequin called 'Ace'. The life-sized figure was soon recognised as a vital member of the Mercedes team and even got his own pink Ventura T-shirt. Elsewhere, Hodgkinson instructed his aero staff to run CFD simulations to analyse the oxygen flow through the device, while Oxford Optronix managed to deliver a self-calibrating analyser in just nine days. 'Two days after we got approval for the Mark I,' Shipley says, 'we got the new version approved too.' The new design slashed the machine's oxygen demands by 70 per cent.

It was now 29 March. The UCL-Ventura team had designed, tested and perfected two devices in 12 days. As the devices were backed by the MHRA, the UK government didn't hesitate to put in an order for 10,000 machines.

Mercedes now had an entirely new problem: how were they going to make them all in just two weeks?

Then disaster struck. Hodgkinson caught Covid-19.

* * *

The rest of the motorsport world had also lent its abilities to the fight against the pandemic. Car manufacturers became mask-makers. Roush Fenway Racing, one of NASCAR's most successful teams, designed what team operations director Tommy Wheeler termed a 'glorified salad bar, sneeze guard type of transport box' to create a barrier between healthcare workers and patients being moved around a hospital. Wheeler and colleagues produced 150 and delivered them to medical centres in five US states. And in Concord, North Carolina, NASCAR's R&D

Center was sharing its manufacturing prowess with another project entirely.

'We built female heads,' says Eric Jacuzzi. 'In the early days, everything was going haywire, so we were doing a bunch of research on the improvised personal protective equipment being used. People were just wearing masks like a bandana over their face, so Wake Forest University asked us if we could get them these ... hollow heads.' The Wake Forest plan was simple: take a head, put a mask on it, and fire Covid-sized particles at its face. 'Basically, you hook a hose up to it,' Jacuzzi says, poking inside a life-size doll head to show me its connections, 'and then they would put a mask on it, or an improvised covering, and measure how much particulate matter got through it. With that, the doctors at Wake Forest could make recommendations intelligently about whether you were actually protecting yourself.' Thanks in part to NASCAR, scientists quickly realised that wearing masks was incredibly effective at curbing the spread of the virus.

Back in the UK, as Mercedes was working flat-out with the UCL professors, the rest of the Formula One teams weren't ignoring the situation. Red Bull, Racing Point, Haas, McLaren, Renault and Williams had banded together with Mercedes to form 'Project Pitlane' and do their bit for the UK government's Ventilator Challenge. This saw teams combine forces and collaborate with industry on an unprecedented scale.

Not all the groups formed by Project Pitlane were successful. Both the Renault and Red Bull teams were involved in one effort, BlueSky, to create a new ventilator for the NHS. Unfortunately, as Shipley had feared, delivering a machine from scratch was unrealistic given the timeframe. After initially placing an order with the BlueSky group for thousands, the government cancelled; clinicians were worried the machine wasn't complicated enough to switch settings easily. Ventilators were just too nuanced to invent in a month, even for F1.

McLaren, as part of the Ventilator Challenge UK Consortium, took a different approach. Rather than come up with an untested design, the Woking team worked in a huge collaboration to manufacture two existing ventilator devices, designed by Penlon and Smiths respectively. Ford took the lead with Penlon; GKN Aerospace and Rolls-Royce with Smiths.

As part of the consortium, McLaren reverse-engineered electronics boards for the Smiths machines, finding alternatives when required parts weren't available. Williams and Renault also worked on prototyping the Smiths designs, then analysed Smiths' production line so other manufacturers could replicate it. The Penlon machine wasn't in line with the government's specifications, so McLaren used its CFD simulators to model its air pathways, identifying what changes needed to be made. In three weeks, the Penlon ESO2 was granted MHRA approval.

McLaren then volunteered to coordinate the consortium's construction needs. 'Managing the supply chain was the toughest part of the project,' wrote McLaren director of innovation Mark Mathieson on the company's website. 'To put it simply, if we couldn't get the parts, then we couldn't make the ventilators.'

In four weeks, McLaren found itself managing a supply network for both Penlon and Smiths, sourcing 10 million parts from 117 suppliers. If a bit was too complicated to order quickly, the McLaren machine shop fabricated it. Thirty-eight staff members worked three shifts, operating the machines almost 24 hours a day for 10 weeks. To keep the production line moving in a socially distanced environment, McLaren engineers stuck whistles to their milling machines' air blasts, which released a shrill toot when a component was finished and the machine could be used again. Consumer goods giant Unilever provided the McLaren machinists with personal protective equipment and food; Hilton Hotels gave them a nearby bed for the night. In three months, McLaren staff created 113,506 individual parts; none of them had a single flaw.

As with Mercedes, the McLaren engineers looked to identify issues before they became problems. Realising the ventilators needed to rest on trolleys – little cabinets with casters – the team gave up its 'build bay', home of prototype supercars, and assigned 35 members of staff to design and build high-grade medical wheelie tables. By the end of the Ventilator Challenge, they had produced 15,000 trolleys for hospitals across the UK.

In 15 weeks, the Ventilator Challenge UK Consortium produced 13,437 ventilators – more than double the number of machines the UK had available at the start of the pandemic. It had been the largest reactive manufacturing project in the UK since the Second World War, involving 3,500 people from hundreds of major companies, delivering the equivalent of 20 years of normal production. At its peak, the Consortium was producing a ventilator every 88 seconds. As McLaren production director Piers Thynne stated, it was: 'A truly astonishing feat given the complexity and number of parts manufactured.'

But while the Ventilator Challenge was still in its infancy, Mercedes were already scrambling to meet the Easter deadline for the UCL–Ventura project. Would they be on time?

★ ★ ★

With Hodgkinson out of action, colleague Pierre Godof took over. The task in front of him was monumental: Mercedes needed to move from making two engines a week to making thousands of CPAP devices a day. A few weeks earlier, most of the team had never even *heard* of a CPAP. 'This is where Mercedes were just phenomenal,' says Shipley, her voice wavering a little in awe. 'I still feel quite emotional about it. At the time, we were so *in it*, so focused, we didn't take stock. All of the people involved were just amazing.'

Instead of starting from scratch, Mercedes looked for a similar assembly line in their own factory that they could repurpose. The closest thing they could find to a CPAP machine was a Formula One car's fuel pump. Soon, they had copied the line across the factory, paying around £50,000 for new machine tools to allow the parts to be made. By now, the UK had entered Covid-19 lockdown, and the Mercedes factory had closed to all but essential workers. But, as news of the UCL-Ventura project spread, the team found itself inundated with colleagues volunteering to return. To speed up production even further, the UK's Manufacturing Technology Centre provided the team with two Automated Guided Vehicles (AGVs). These were driverless forklift robots, built to help shift parts around a factory and are just as cool as they sound. The two AGVs were soon put to work, covering 20km a day to ensure the

Mercedes production lines flowed. At first, the machines gave off a '*wurp, wurp*' alarm as they trundled along. Within a day, the F1 engineers had hacked the robots, programming them to zoom about to the Imperial March from *Star Wars*. At their peak, the team's 12 production lines were producing 1,200 CPAP machines a day.

As with McLaren, the UCL-Ventura team found sourcing the right parts was incredibly tough. The international borders had shut so Mercedes couldn't call on their suppliers in Europe, while most factories in the UK had closed because of the Covid-19 lockdown. Hodgkinson, recovered and back at work, found that everyone was willing to help. 'We'd explain what the project was for and, in six hours, a factory would be back up and running, chopping plastic to send to us.' Soon, Mercedes was receiving three deliveries a day from its suppliers to keep the production lines running. When the team's supply chain was stretched to its limit, Baker again used his motorsport contacts, asking LifeRacing to design and manufacture 8,000 analysers based on automotive oxygen sensors (once again needing MHRA approval). It decided to call the device ThanOx and the company it spawned Thanos Medical. Even a Marvel supervillain was helping to fight the pandemic, albeit from the unlikely locale of a trading estate outside Basildon.

The UK government was aware of how well the project was proceeding from the start. Playing an unsung role here was David Lomas, UCL's health vice provost and a respiratory consultant at UCLH; using his lobbying skill and contacts, Lomas had made sure UCL-Ventura was on the government's agenda as well as the Ventilator Challenge itself. Every day, the Cabinet Office's Lord Agnew would contact Mercedes to ask for an update. Sometimes, more senior figures got in touch. At one point, Hodgkinson was standing next to Shipley while she was finishing a phone call. Suddenly she paused, turned, and asked him to put together a presentation on the project for 'someone in government'. 'I didn't get on it immediately, I was finishing a job I was doing,' Hodgkinson later recalled. 'And she said "No, can you do it now, like, now?" So I started putting it together. And then she said, phone to her ear, "I've got Michael Gove on the phone, he's in the office with Boris [Johnson, the Prime Minister] and we need it *now*!""* Hodgkinson bashed out his presentation in 'about ninety seconds' and passed it over to Shipley, who took a photo of it on her phone and sent it via WhatsApp.

The 10,000th UCL-Ventura CPAP machine was delivered to the NHS on 15 April 2020. It had been just 30 days from when Baker, Singer and Shipley had met in the university's staff common room. By the time the first wave of the pandemic hit its UK peak, machines built at Mercedes were available in more than 60 hospitals across the country. The team's superhuman effort had saved countless lives.

'If you look back now at the intensive care audit data from wave one [of the pandemic],' Shipley says, 'mortality rates dropped 26 per cent and there was a reduction in the use of ventilators by 23 per cent for equally sick patients.' While this wasn't down to CPAP machines alone, their role has been recognised at UCLH and beyond. 'Luckily, with the lockdown it never came close to anything like the predicted demand on

---

* Shipley recalls it wasn't quite this dramatic; she was on the phone to Lomas, who was on the phone to a member of the House of Lords, who was on the phone to Gove. Either way, it's pretty cool.

the NHS or ventilators,' Singer later recalled. 'But certainly in London we were teetering on the edge. And the use of CPAP saved us.'

The team's efforts didn't end that April. As the 2020–21 winter set in, the UK faced a second Covid-19 wave, which placed even greater strain on hospitals. Yet again, the data shows the team's decision to focus on CPAP machines was the right one. 'We were even closer to the edge in wave two,' Shipley recalls. 'But even so, we peaked at around 4,000 people on ventilators. Availability of staff was the issue – normally you would have one-to-one nurse-to-patient ratios, but that was diluted down to one-to-three or even one-to-four in some cases. The idea that we could have looked after thirty thousand people on ventilators is a fantasy.'

Quite simply, Formula One had saved the NHS. 'For me, it shows what you can do with medical devices if you have the capability [of an F1 team] on board,' Shipley says. 'There's quite a bit of difference, in all sorts of ways, between the NHS and Formula One. Not least the financial investment! But it really brings home the level of precision engineering and capability they have. In a utopian world, what could we do with our healthcare system if this was the way it operated?'

It's a question that remains, for now, unanswered. Mercedes has also learnt from the project. The company's processes have been streamlined, inspired by some of the workflows they've seen in hospitals. And the two AVG robots are staying put, *Star Wars* tunes and all.

This isn't the end of the UCL-Ventura story. Realising that human life comes before profit, the project has since made its designs and manufacturing instructions available for free to anyone who can get medical approval, will follow their patterns and pledges not to make an excess profit with the device. 'So far, there's been one-thousand-nine-hundred approved licences [to make the UCL-Ventura CPAP] across one-hundred-and-five countries,' Shipley says. 'There are now twenty teams that are manufacturing them in volume, and some have regulatory approval, too, such as in Peru, Paraguay, South Africa, India and Pakistan.'

Rather than rest on their laurels, the UCL-Ventura team have continued to work together. There are weekly phone calls between the university and the Brixworth engineers. They have created training packages to help people get through the technical and clinical aspects of regulatory approval and deployment, and Shipley is liaising with charities to send their devices around the world. 'I just got an email through with a picture of some CPAPs that had cleared customs in Uganda and were on their way to Lacor Hospital,' Shipley says. 'I sent it through to the team at Mercedes.'

The incredible work has already gained some of the recognition it deserves. In October 2020, Baker was awarded an MBE for services to healthcare in the UK and abroad. Yet the true reward for the UCL-Ventura team is knowing they made a difference. When racing finally started, Mercedes romped home to victory. The team won the 2020 Formula One constructors' championship title and Lewis Hamilton claimed a record seventh world title. But many of the team consider their delivery of the CPAP devices the real victory that year. 'We'll never, ever forget it,' said Cowell. 'It's a real privilege to work with people in this business. You don't have to say anything. You just have to provide people the space, the tools, the direction, the target, and you can stand back and watch the magic happen.'

For Shipley, the project will also stay long in the memory. 'This is going to be a real career high,' she says. 'There's a real sense of privilege that we were able to work together and contribute. You don't get many opportunities to do something like this.' She pauses ruefully. 'Well, I *hope* we don't ever have a year like that again.'

I hope so, too. But the Covid-19 pandemic wasn't the first time Formula One had to learn how to save lives. Fifty years ago, it was the deadliest sport on Earth.

CHAPTER NINE

Matters of Life and Death

In 2008, a triptych by Scottish artist Jack Vettriano was unveiled at the Hôtel de Paris in Monaco. Titled *Tension, Timing & Triumph*, the vignettes showed Sir Jackie Stewart's victory at the 1971 Monaco Grand Prix: his wife Helen watching his lap times on a stopwatch; the imperious cool of Stewart, already a world champion, marching to his car; the couple's embrace at his victory. Or, perhaps, an embrace thankful that Stewart had survived.

A few years after that Monaco triumph, Stewart was in the closing stages of the 1973 season, driving for the dominant Tyrrell team. He'd already won the championship and had secretly sworn the final race – the United States Grand Prix at Watkins Glen – would be his 100th and last. The plan was to hand the lead driver baton to his teammate and friend, 29-year-old François Cevert. It was a secret even Helen didn't know. On the morning of Saturday, 6 October, the cars began qualifying. Eager to prove himself to the team, Cevert was hunting for pole, slipping through a fast series of turns on the far side of the track known as the Esses. Handsome, charming and talented, Cevert was the model of a relentless racer. Taking the turn in third gear, he ran on to the left kerb, slid across the track and hit the right guardrail. This spun the car back across the racing line, this time at a 90-degree angle, where it hit the far barrier at almost 150mph. When the marshals reached the scene, they didn't even try to get Cevert out of the car: he had been almost decapitated. Stewart arrived, saw his friend, teammate and successor dead, and drove back to the pits. When Lotus' Colin Chapman saw Stewart climb out, the story's tragedy was writ large among tears in the Scot's eyes.

Cevert was gone. Stewart was done. Tyrrell withdrew from the race and Stewart walked away from the sport with 99 starts to his name. He valued his life more than a round number.

Stewart had been a racing safety advocate long before his decision to leave the sport. At Spa in Belgium in 1966, he'd skidded off the track at 165mph, crashed through a telephone pole and shed and ended up trapped in a farmer's outbuilding. The steering column had buckled and pinned his leg, and the fuel tank's contents poured into the cockpit, drenching him in gallons of easily ignitable fluid inches away from heat and flame. There were no jaws of life – the tools firemen use to cut people from cars – available. 'Steering wheels weren't removable in those days,' says Mark Gallagher, former Formula One executive with Jordan, Jaguar and Cosworth, and an expert on the history of the sport. 'He was trapped and he was convinced he was going to burn to death.'

Stewart's life was saved by fellow drivers Graham Hill and Bob Bondurant, who hauled him out of the wrecked car. But Stewart's woes were only beginning. He needed medical attention but there was none to be found. Instead, he was left in the back of a pick-up truck until an ambulance could find the crash and take him back to the circuit's first aid centre. There, Stewart's stretcher was dumped on a messy, litter-strewn floor amid empty cans and cigarette butts until another ambulance could come and take him to hospital. Unbelievably, that ambulance driver got lost on the way.

The incident at Spa haunted Stewart for the rest of his career, so much so that he hired a private doctor to accompany him at all future races. 'He would also keep a spanner in the cockpit at his side,' Gallagher adds, 'so that if there was an accident he could undo the steering wheel.'

After Spa, Stewart focused his star power on improving safety. He campaigned for mandatory seat belts and full-face helmets for drivers, better track organisation, run-off areas, fire crews and medical facilities. It was a move that ruffled feathers. 'I would have been a more popular world champion if I had always said what people wanted to hear,' Stewart later recalled. 'I might have been dead, but definitely more popular.'

Stewart's safety estimate was brutal: any driver who raced for five years had a two-in-three chance of being killed. 'Jackie had a quite short career in racing,' Gallagher says. 'But, during it, he and Helen worked out he personally knew fifty-six drivers who lost their lives.' They included fellow Scot and double world champion Jim Clark, killed at Hockenheim in Germany in 1968, and, in 1970, the only posthumous world champion, Jochen Rindt, who was killed in practice during the Italian Grand Prix with four races still to go in the season.*

Rindt's death only underlined the need for better safety equipment and for stark conversations about using it. Seat belts were new arrivals in Formula One but, although most teams had the belts installed, only Stewart and a few others wore them properly. Earlier that year, driver Piers Courage had burnt to death after being trapped in his car and most

* This tally doesn't include drivers killed off track, most notably Hill, who died in a plane crash in 1975 shortly after his retirement from racing.

drivers wanted to make sure they could escape quickly; the prevailing theory was that it was better to be thrown clear in an accident rather than be pinned to your seat. Rindt himself wore the belt across his chest loose, and had demanded his mechanics remove the crotch strap, designed to prevent a driver slipping down in their seat, or 'submarining', during an accident. Tragically, the decision cost him his life. Rindt suffered a brake system failure and, largely thanks to his Lotus-72's instability due to its lack of wings (see Chapter Five), Rindt was sent spinning off track. He crashed into a barrier, where the impact caused him to submarine under the chest's straps. Horrifically, he was garrotted by his own loose seatbelt.*

Today, this couldn't happen. Drivers use a six-point safety harness (two leg straps, two shoulder straps, two pelvic straps, all released with a twist of a hand) to prevent the driver submarining in the same way. But Rindt's death is indicative of the safety culture at the time. And, although we might think of a seat belt as a fundamental safety feature, keep in mind that NASCAR only made six-point seat belts mandatory in 2007.

In Formula One, the deaths continued throughout the 1970s. Revlon heir Peter Revson. Helmuth Koinigg, also at Watkins Glen, in an accident virtually identical to Cevert's. Mark Donohue, along with marshal Manfred Schaller, killed by debris thrown from Donohue's car. Tom Pryce, killed in a

* Rindt did not die immediately; he was collected in an ambulance Bernie Ecclestone described as 'more like a pick-up truck' and taken to hospital. Despite leaving the track after Rindt, Ecclestone arrived at the emergency ward before Rindt: the ambulance had gone to the wrong hospital.

collision with another marshal, teenager Frederik Jansen van Vuuren, as the latter rushed across track to put out a fire. The names kept coming. So, too, did accidents resulting in life-altering injuries – the most famous of which is the blaze that almost killed world champion Niki Lauda, leaving the Austrian with horrific lung scarring and burns across his face and body for life.

Too many people were dying or being maimed in the name of entertainment. And Formula One knew it had to stop.

* * *

Despite appearances, people weren't blind to the danger of the track. Throughout the 1960s and 1970s, Formula One had introduced a host of regulations for safety. These included roll bars, seat belt rules, fire protection and filters, electric circuit breakers, reverse gears and safety structures. Drivers began to wear protective equipment (with standards for fire resistance and helmets) and the FIA took over safety from local bodies, introducing flag signals and staggered grids. Tragically, these measures weren't enough, but it would be disingenuous to suggest motorsport didn't try to look after its drivers. In fact, it was racing that led to some of the most basic protections we take for granted today.

In the early twentieth century, the 'Fastest Girl on Earth' and 'Champion Lady Motorist of the World' was Dorothy Levitt. A jeweller's daughter from Hackney in east London, little is known about Levitt's early life save that she was an expert horse rider. In 1902, aged 20, she arrived at the offices of British motor pioneer Selwyn Edge and became his personal secretary. In Levitt, Edge spotted a potential marketing opportunity and decided to make her the first female British woman racing driver: an East End girl who could beat the toffs at their own game. Soon after, Levitt was sent to Paris to master the art of driving at speed.

Free-spirited, ferociously smart and a renowned beauty, Levitt excelled. First, she piloted motorboats, setting the world's first Water Speed Record of 19.2mph in July 1903 (although, as owner of the boat, Edge's name went on the trophy). Then she moved to the road, winning her class at the Southport Speed Trial, driving from London to Liverpool and back again in two days, and setting the women's land-speed record at the Brighton Speed Trials. The next year, she broke the record again with a speed of 90.88mph. 'Had near escape as front part of bonnet worked loose,' her diary entry for the day noted. 'Had I not pulled up in time, might have blown back and beheaded me.'

It was this cavalier attitude that made Levitt the toast of London. Rather than hide her sex, she revelled in it, wearing deliberately feminine racing outfits with a coordinated hat and veil. Her constant companion was Dodo, a yapping hairball Pomeranian given as a gift in France, who Levitt smuggled into England drugged and hidden in her car's repair box. (At the 1904 Hereford Thousand Miles Trial, her fellow competitors, fed up of Dodo barking at them constantly, turned up with toy dogs strapped to their cars.) Alluring and charming in society – she would later teach Queen Alexandra and the royal princesses to drive – Levitt also had a temper. In November 1903, she was stopped for speeding through Hyde Park. Rather than apologise, she told the officer that she 'would like to drive over every policeman and wished she had run over the sergeant and killed him'. This resulted in a magistrate fine of £5 2s (about £650 today). She also learned how to fly, becoming one of the first women to pilot an aeroplane.

Levitt's link to safety came in 1909 in her book *The Woman and the Car: A Chatty Little Handbook for all Women Who Motor or Who Want to Motor.* Based on her newspaper column, the guide was peppered with support and advice for girls who wanted to be recognised as equals on the road. It included how to start a car, what to wear ('Under no circumstance wear lace or fluffy adjuncts to your toilet'), through to self-defence ('If you are going to drive alone in the highways and byways it might be advisable to carry a small revolver. I have an automatic Colt and find it very easy to handle … '). Its best piece of advice was a simple tip that we find obvious: to keep a pocket mirror 'handy, not only for personal use, but to occasionally hold up to see what is behind you'. Dorothy Levitt had popularised the rear-view mirror, seven years before it was adopted by manufacturers.*

The crash helmet is another motorsport invention, emerging from motorbike races at Brooklands before the First World War. The medical officer for the track, Eric Gardner, noticed that he was seeing at least one racer with a serious head injury every two weeks, and commissioned canvas and shellac helmets to protect his charges from a hard impact or glancing blow. The move faced an immediate backlash, but he managed to persuade the organisers of the 1914 Isle of Man TT to make helmets compulsory for the race. When one of the riders crashed into a gate and the helmet saved his life, the grumbling stopped.

---

* Despite her exploits, Levitt died in obscurity in 1922 from a morphine overdose while suffering from heart disease and measles. She was 40 years old.

But it would take a far more famous death to make helmets compulsory for all. In 1935, Colonel Thomas Edward Lawrence was riding his motorbike close to his home in Dorset when he swerved to avoid two boys out cycling. Thrown over his handlebars without a helmet, Lawrence died six days later despite the efforts of his neurosurgeon, Australian army veteran Sir Hugh Cairns. The death of the fabled 'Lawrence of Arabia' troubled the medic and he began to focus his research on how to prevent motorbike fatalities.

When the Second World War broke out, Cairns became a consultant to the army and soon revolutionised military healthcare. He authorised the first clinical trials of penicillin, thus playing a crucial role in the discovery of antibiotics. He created mobile neurosurgical units to aid troops on the front lines, which led to the British Army having the lowest death rate from head injuries of any combatant in the war. And, finally, he was able to do something to prevent future motorcyclists from meeting Lawrence's fate. In 1941, Cairns published a paper in the *British Medical Journal* looking at deaths from dispatch riders. 'During the first twenty-one months of the war,' he began, 'the number of motor-cyclists and pillion passengers killed on the road was 2,279 … the frequency of head injuries was high and in a number of cases the fatal outcome might have been avoided if adequate protection for the head had been worn.' Conversely, Cairns reported he had only seen seven injuries in motorcyclists wearing crash helmets – and not one had been fatal.

Thanks to Cairns, crash helmets became compulsory for British dispatch riders. In 1946, he published a follow-up paper, this time with a graph showing the difference in casualties once helmets were introduced. Despite more men enlisting, and more people on motorbikes, the number of fatalities had more than halved.* Mass-produced helmets only became available in 1954, with the Bell 500, a fibreglass helm that wisely extended protection around the sides and back of the head. Soon, helmets became a fixture of races around the world, although it wouldn't be until 1973, the same year Stewart quit motorsport, that they became compulsory on UK roads. Gardner's invention – and Cairns' persistence – saves thousands of lives a year.

Today, crash helmets have evolved into a state-of the-art protection. A modern helmet does not merely shield the back and sides of a head but is fully enclosed with a visor (which includes several layers that can be torn off if they get dirty during a race). The best way to think of a crash helmet is like a hard-boiled egg, with the driver as the gooey yoke in the centre. On the outside, a rigid outer shell offers a first line of protection, typically made of tightly woven Kevlar fibres. This is designed to protect from flying sharp objects or massive blunt trauma. Even so, something impacting a crash helmet will still shake the head and cause injury. So, beneath this shell is a softer layer (the egg white in our example), which deforms to absorb the kinetic energy of sudden movement. The current FIA standard for a helmet requires it to withstand a 225g projectile fired at 250km/h. It also specifies the limits of how much energy it needs to absorb to protect the head. And, of

* Cairns H. *Crash Helmets*. *BMJ* 1946, 2:320 (DOI: 10.1136/bmj.2.4470.322). The title of this chapter comes from the classic Powell and Pressburger film *A Matter of Life and Death*, in which a motorcyclist is killed while not wearing a helmet.

course, the helmet needs to be fire resistant: the FIA requires it to self-extinguish after being exposed to a 790°C flame.

Virtually all helmet standards we follow today stem from racing. In 1956, sports car racer Pete Snell died of head trauma after the crash helmet he was wearing failed. The following year, his friends, along with scientists and doctors, set up the Snell Memorial Foundation to create a set of regulations for all types of safety helmet. Today, you can find Snell certification on helmets used for everything from kids' bicycles to horse riding, skiing and karting. If you've ever done one of these activities, chances are you have Lawrence of Arabia and Pete Snell to thank for doing it safely.

But while you might live in a country where these standards are mandated, many do not. Of the 300 million motorbikes and mopeds in the world around 80 per cent are in Asia, often in countries without adequate regulations, contributing to the 1.35 million road fatalities globally every year. To tackle the problem, the FIA tasked its safety engineers with designing a low-cost, lightweight motorbike helmet suitable for the tropics that could pass UN safety standards. In 2019, the team did it, manufacturing 3,000 helmets for field tests in India, Jamaica and Tanzania. The Covid-19 pandemic disrupted the programme but it's hoped that soon, thanks to racing, every motorcyclist in the world will be able to get adequate head protection for as little as $10.

★ ★ ★

In 1978, Brabham team boss Bernie Ecclestone drove along Whitechapel Road to pay an early evening visit to The London

Hospital. There, he made his way up to the neurosciences department and called in to the office of 49-year-old surgeon Sid Watkins. A former army doctor who had worked in the UK and US, Watkins was a passionate fan of motorsport. He had competed in one stage of the West African Rally in 1955, and had served as a doctor at Silverstone and Watkins Glen. But Ecclestone had a greater role in mind. He knew it was no longer enough just to have a doctor on standby at races and rely on each track to provide barely adequate medical care; for Formula One to thrive, the sport needed to take charge of safety itself. In Watkins, Ecclestone believed he had found the perfect man to do it.

Watkins agreed. 'Professor Sid' had arrived in Formula One as its first safety and medical delegate.

Watkins is, rightly, held in almost reverential awe by the racing fraternity. For 26 years, he constantly looked for ways to tighten and improve facilities, training and procedures. It is a sad truth that everyone in motorsport knows someone who was killed in an accident, from the 1960s through to the sport's most recent death – and only fatality of the twenty-first century – Jules Bianchi, who died in 2015 from injuries sustained at the 2014 Japanese Grand Prix. But everyone also knows someone who wouldn't be alive today if it wasn't for Watkins. On track, he saved the lives of drivers including Gerhard Berger, Érik Comas, Martin Donnelly and Didier Pironi. At the 1995 Australian Grand Prix, he and the local medical team twice restarted the heart of future world champion Mika Häkkinen and performed a cricothyroidotomy on track, creating an airway through the Finn's neck as he waited for the ambulance to come. During the off-season, Watkins' surgical skills pioneered work in brain and spine stimulation as well as pain management and Parkinson's disease. During the 1980s, he was instrumental in establishing the London Air Ambulance and his hospital's emergency brain-scanning unit – an innovation that's been copied around the world.

The surgeon soon realised there were basic, fundamental problems in Formula One. Medical care was highly variable and in many cases completely detached from reality. 'Medical centres, their equipment, staff, training, competence and attitudes were frequently inadequate,' he wrote in his book *Life at the Limit: Triumph and Tragedy in Formula One.* At his first race, the Swedish Grand Prix, the medical centre was a caravan with a large hut 'to house extra casualties', and no helicopter was available during practice because it was 'not dangerous compared to the race' – a fact that blithely ignored the statistics. At Brands Hatch, the medical centre was underneath the grandstand with no direct access from the circuit; when Watkins arrived, the two ambulancemen were drinking beer and had lost the key to switch on the oxygen supply. Hockenheim didn't even *have* a medical centre, although Watkins immediately saw the potential of the track's fast Porsche intervention cars, driven by professional racers to get medics to an accident quickly. 'Everyone talked about safety and did a lot of things to try and improve it,' Gallagher says. 'But they needed someone to actually implement the changes that needed to happen – someone who knew exactly what it was that was killing people. That was Sid.'

Watkins faced stiff opposition from some tracks. At Hockenheim in Germany, Watkins was denied access to race control – prompting Ecclestone to threaten to stand in front of the grid and order the drivers out of their cars if the medic wasn't allowed inside. But the track Watkins always hated the most was Monza in Italy, his 'personal nightmare of the year', with Watkins writing in the 1990s that 'despite my congratulating the authorities there on having the worst medical centre in the world, little has been done to make it more than the basic standard acceptable'. This included having the right people on hand; once, Watkins turned up at Monza to interview the medical staff and check their skill sets. Among the emergency team he found a physiotherapist and a nutritionist. 'What help are you going to be?' Watkins asked. 'Are you going to give them a massage and a nice meal?'

Monza was also the track that showed Watkins the sheer scale of the task he faced. On 10 September 1978, the Italian Grand Prix began with a disaster. The lights went green too early causing 10 of the 24 cars to smash into each other as the pack funnelled. While most drivers escaped with minor injuries, the Lotus of Ronnie Peterson was pushed hard into the barriers, where it burst into flames. Drivers James Hunt, Clay Regazzoni and Patrick Depailler were able to pull Peterson free, but Watkins was unable to get to the casualty: the carabinieri had formed a line across the track and inadvertently blocked all medical help. 'That's how basic things were, how wrong things were,' Gallagher says. 'The head of medical services for Formula One, with a pass around his neck, couldn't get to the accident.'

Despite the hold-up, Peterson survived the initial crash: his legs were smashed in 27 places, but it wasn't a fatal injury. Unfortunately, that night he developed a fat embolism that caused kidney failure, and died the next morning. 'He should have been fine,' Gallagher says. 'But he wasn't in the right hospital, with the right people, giving him the right attention. And that's when Sid realised he couldn't just be a doctor, he had to take a holistic view. First, let's make sure the accident doesn't happen. Second, if it does, make sure the human being doesn't get injured. And third, if they do get injured, make sure they're treated within seconds, by the right people and then flown to a trauma facility that has the ability to protect them.'

After Monza, Watkins insisted the medical car follow the drivers for the first lap. Soon after, it was decided that Watkins wouldn't merely be Formula One's doctor; he would also decide race safety procedures. The era of pot-luck medical facilities was over.

Improving safety was never going to be an overnight fix. Culture and procedures needed to change, and this took time. In 1980, the same year Formula One established its Medical Commission with Watkins as president, Depailler died in a test at Hockenheim. And, in 1982, two more deaths hit the sport. On 8 May 1982, the hero of French Canadians, Gilles Villeneuve, was thrown from his car in a crash at Zolder, Belgium, still strapped to his seat but somehow without his helmet. The first medic was with him in 35 seconds; Watkins in little over two minutes. But Villeneuve's neck had been fractured, and there was little either doctor could do. A little over a month later, with Canada still mourning their national hero, Formula One arrived at the Circuit Île Notre-Dame in Montreal. There, rookie driver Riccardo Paletti was also killed. In 1986, yet another driver, Elio de Angelis, lost his life during testing; at the time, tests didn't follow the same rigorous safety procedures as a race. And yet, gradually, Professor Sid's work began to change outcomes.

'What he did was reverse-engineer why people were killed in a car,' Gallagher explains. 'He started with the person who was dead and asked what killed him. Usually, that was because the car stopped so suddenly while your internal organs don't. They decelerate within the skull or the torso, resulting in horrific internal injuries. You weren't dying from a car crash; it was from crashing in a way that doesn't absorb any energy, so it goes all into your legs and torso.' Throughout the 1980s, the changes focused on solving the problem: crush zones, designs of the cockpit to ensure driver protection, changing survival cell material from aluminium to carbon fibre, the integrity of crash barriers, the size of run-off areas, the protection of crew in the pit lanes. Every year, the safety net grew.

Another major focus was fire. 'Think about how fire works,' Gallagher explains. 'You need fuel. So, they developed self-sealing fuel lines, which sealed themselves after a cut. And they used Kevlar and carbon fibre in fuel tanks, so they are a fully deformable structure that's basically bulletproof, nothing will go through them. The other day I was laughing with Martin Brundle, who had got to drive one of Sterling Moss' old Formula One cars at Silverstone. And Martin said he suddenly realised he was sitting in a car where the only thing supposedly protecting him in a crash was two fuel tanks! When you look back at the 1960s and 1970s, it seems so obvious now. What on Earth did they think was going to happen?'

The results of these changes are clear from the statistics. According to Watkins, between 1963 and 1982, there were 263 Formula One races with an estimated 1.3 million km travelled. During them, there were 668 accidents, 106 severe driver injuries and 19 fatalities (including one official and six spectators).* From 1983 to 1996, there were 208 races, around 1.1 million km travelled, and 771 accidents – but only five severe injuries and two deaths.

Their names were Roland Ratzenberger and Ayrton Senna.

★ ★ ★

As I said in the introduction to this book, I'll never forget the day I heard of Ayrton Senna's death. It is a permanent scar, a painful mark seared upon my childhood. Its effect on the world of motorsport was just as profound.

The 1994 San Marino Grand Prix is arguably the blackest event in Formula One history. San Marino is a tiny country, basically three towers at the top of a mountain, so it doesn't have a racetrack. As such, the race it gave its name to was held at Imola in Italy, about 60 miles to the north west, halfway between Bologna and Ravenna. The meet did not begin well. On Friday, 29 April, Senna's protégé, Rubens Barrichello, hit a kerb at 140mph, launching his Jordan into the air, where it struck the top of a tyre barrier. The car landed upside down, knocking the driver unconscious with an impact measured at 95*g* (in other words, 95 times the normal gravitational acceleration on Earth). Barrichello almost choked to death, but Watkins arrived and saved his life. Ultimately, the Brazilian survived with minor injuries. Racing continued.

* These figures are only for official Formula One events; most drivers raced in several series and non-championship races. They also do not include accidents during testing.

The next day was qualifying. Simtek's Roland Ratzenberger, unable to control his car with a damaged front wing, crashed at the Villeneuve kink, hitting the concrete barrier there head-on. The car's survival cell wasn't broken but the impact resulted in a basal skull fracture (a break in the bone at the base of the skull). He was airlifted to hospital but died of his injuries.

The news affected Senna greatly. A personal friend of Watkins, the Brazilian broke down and cried on the doctor's shoulder at the news of Ratzenberger's death. Already a three-time world champion, Senna was the sport's talisman, a mercurial genius who infuriated and infatuated friends and rivals alike as he battled Alain Prost and Nigel Mansell. Watkins tried to persuade his friend to follow in Jackie Stewart's footsteps, to get up and walk away from the sport and go fishing with him instead. Senna declined. 'Sid, there are certain things over which we have no control. I cannot quit, I have to go on.'

Sunday was race day. It started badly. One of the Benettons stalled on the grid, resulting in a Lotus ploughing into its back and sending car parts flying over the safety fencing, where they injured a police officer and eight spectators. This led to a safety car, an Opel Vectra, coming out to lead the pack as the debris was cleared. But the Opel's brakes had started fading and the safety driver was forced to reduce speed, travelling too slowly to warm up the competitors' tyres. Senna, on pole, pulled up alongside several times, urging the car to go faster. The scene was set for one of the worst tragedies in motorsport history.

The race restarted on lap five. Two laps later, Senna was leading but, with his cold tyres, he was unable to take the turn at Tamburello. Instead, he continued straight, hitting a concrete barrier lining the circuit at somewhere around 131mph. The force caused Senna's front right wheel and its assembly to rip off and fly into the cockpit, where it broke through his helmet, crushed his skull and impaled him just above the eye. As marshals rushed to Senna's aid, there was a final, unnatural roll of the head, then only stillness.

Watkins knew instantly from Senna's pupils that the driver had suffered a massive brain injury. Even so, the doctor did his duty, performing a tracheotomy, clearing respiratory passages, stemming the flow of blood and immobilising Senna's spine. The Brazilian was flown to hospital within 28 minutes of the crash, but he was already gone. When his heart stopped a few hours later, it was decided not to restart it. He was 34 years old. When they searched his car, they found an Austrian flag; Senna had intended to fly it in honour of Ratzenberger.

A few weeks later, as the world mourned the passing of a generational talent, a final pivotal accident occurred at the Monaco Grand Prix, this time to Karl Wendlinger. Thanks to immediate, excellent medical attention – proper spinal and neck splints, clear airways, oxygenation, mannitol infusions to reduce brain swelling and an immediate transfer to a state-of-the-art neurological unit – the driver made a full recovery. But four serious accidents in less than a month – Barrichello, Ratzenberger, Senna and Wendlinger – had shaken Formula One to the core.

* * *

In 2001, NASCAR would face a similar turning point. 'There's a machismo culture in NASCAR,' says physicist and NASCAR expert Diandra Leslie-Pelecky. 'The biggest barrier to implementing safety is usually the drivers. They are mostly young men, aged twenty-five to thirty-five, who think they're invincible. They take a lot of risks and, for a long time, NASCAR was just corralling them. Back in the 1960s and 1970s, they used to do tyre tests by getting drivers to run around a track at speed, then throwing up tacks and nails to make the tyres blow. People volunteered for it! But then, in 2001, there were four deaths over a ten-month period because the cars were not safe enough and the tracks were not safe enough.'

The first was 19-year-old Adam Petty, America's first fourth-generation sportsman. His great-grandfather Lee won the 1959 Daytona 500; his grandfather Richard 'The King' Petty holds almost every NASCAR record to this day; his father Kyle is an eight-time race winner. On 12 May 2000, Petty was in a practice session at New Hampshire Motor Speedway when his throttle stuck wide open, causing him to hit the outside wall head-on. The scion of an American dynasty was killed instantly from a fractured skull.

Eight weeks later, 30-year-old Kenny Irwin Jr died at the same track, on the same turn, when his car flipped on to its roof. Again, his death was instantaneous from a fractured skull. Then, in October, 35-year-old veteran Tony Roper crashed during a truck race in Texas. Pulled unconscious from his vehicle, he was taken to the in-field medical centre then flown to a hospital in Dallas. He died the next day.

But it was the fourth death that really shocked everyone. On 18 February 2001, the Daytona 500 ran with an estimated 17 million viewers on TV. A year earlier, sport superstar Dale

Earnhardt had criticised new rules on springs and shocks in the cars, leading to NASCAR developing a new aerodynamic package that would cause cars to group close together and allow high-speed passing. Daytona would be the first 500-mile race to use the new design.

The majority of the race went well. On lap 173 of 200, Earnhardt's team held the top two positions, with Earnhardt himself, driving for Richard Childress Racing, in third just behind his son, Dale Jr. Then, at the back of the bunched pack, a crash caused cars to ping-pong off each other, eliminating 18 of the drivers. The race was immediately red-flagged, with Earnhardt chatting to his crew over the radio to criticise the new package. 'If they don't do something to these cars,' he drawled, 'it's gonna end up killing somebody.'

On the final lap, Earnhardt was still in third. Then fourth-placed Sterling Marlin attempted an overtake. Earnhardt made contact with Marlin's car, went off course and crossed in front of fifth-placed Ken Schrader. The two hit, sending both into the retaining wall at around 160mph. Earnhardt hit harder, crashing at an angle of 55 degrees. Across America, spectators watched in horror as the front of his car crumpled, ripping the bonnet loose and causing it to smash repeatedly into Earnhardt's windscreen.

Eventually, the wreck of the two cars skidded to a halt. Schrader, with only minor injuries, pulled himself out of the cab and rushed to check on Earnhardt. He glanced inside the cab, pulled the safety netting down and signalled for the paramedics. A decade later, Schrader confessed he knew Dale Earnhardt was dead from the moment he looked in the car. He just couldn't bring himself to say it for years.

For the third time in 10 months, a NASCAR driver had suffered a fatal basal skull fracture. 'Previous accidents were written off,' Leslie-Pelecky says. 'They'd say, "Oh, it wasn't an experienced driver", or, "Oh, it was a freak accident". None of that applied to Dale Earnhardt. Earnhardt was God. And when God dies, it forces a lot of people to say, "We've got to do something."'

★ ★ ★

The changes both Formula One and NASCAR went through were sweeping. Everything, from procedures, car design and the tracks themselves were overhauled for safety. 'Anyone who works in a sector where safety is key can tell you of a particular incident that changed everything,' Gallagher says. 'For North Sea Oil, it was Piper Alpha. For Network Rail, it was the Hatfield rail disaster. And for Formula One, it was that weekend in San Marino. You can effectively plot the safety of motorsport as a story of before and after San Marino.'

The greatest of these changes has its origins much earlier, with a crash involving a far less star-studded name than Senna or Earnhardt. In 1981, Patrick Jacquemart was testing his Renault Le Car at the Mid-Ohio Sports Car Course when he struck a sandbank. The trauma of the sudden stop resulted in the same injury as Ratzenberger, Petty, Irwin Jr and Earnhardt suffered: a basal skull fracture. Jacquemart was rushed to hospital but was pronounced dead on arrival.

Shortly after, road racer Jim Downing discussed the accident with his brother-in-law, Robert Hubbard, a biomechanical engineer at Michigan State University. Hubbard realised a simple truth. While seat belts protected a driver's chest, there was no equivalent protection for the head and neck. Keen to fix the problem, Hubbard set to work. By 1985, he had prototyped his design and, in 1989, it went through its first crash tests. It was called the Head and Neck Support, or HANS, device.

The HANS device is simple in design. It's a rigid carbon-fibre collar that fits over the shoulders, secured in place by the seat belt. At the back are a pair of tethers, which attach to the driver's helmet. This prevents the neck from flying forwards and hyperextending in a crash, reducing neck tension by more than 70 per cent and avoiding basal skull fracture.

Initially, no one was interested in the design. Frustrated, Hubbard and Downing set up their own company to develop, manufacture and sell the HANS device. Even so, it remained unpopular throughout the 1990s. Earnhardt himself referred to the device as 'that damn noose', while other drivers refused to wear one because it was so uncomfortable.

San Marino changed everything. After Ratzenberger's death, the FIA began to test different options for reducing deceleration injuries to the head and eventually decided the HANS device was the best solution. It was introduced into Formula One in 2003. Two years earlier, following the death of Earnhardt (and a non-NASCAR, and fortunately non-fatal, race accident involving Earnhardt's son Kerry), NASCAR had also mandated the used of the HANS device for its top three series.*

---

*You might be wondering why you don't use a HANS device in your car. The answer's easy: you don't have a thick enough seat belt to keep the yoke in place, and you aren't wearing a crash helmet you can connect to the HANS' tethers.

For NASCAR, perhaps the biggest safety shake-up involved the cars themselves. Before Earnhardt, the chassis (even during the 'Twisted Sister' days) were based on a 1966 Ford Fairlane. Instead, NASCAR began to develop the 'Car of Tomorrow'. The seat was moved four inches towards the centre, the roll cage three inches to the rear. Larger crumple zones were included, as was foam into the car's sides. Even the exhaust was moved, diverting heat away from the driver, while the fuel cell was shrunk and strengthened. In 2003, the sport mandated tethers made of Vectran – a polymer of similar strength to Kevlar but ten times more flexible – to be used to secure any part of the car that could easily detach. Similar tethers had been in use in Formula One to protect from flying wheels since 1999.

'People say that twenty years ago was the greatest time for NASCAR, but they forget what it was like,' Eric Jacuzzi says. 'Today, on the safety side, there's never any hesitation from NASCAR. We will do the right thing, whatever it takes. A prime example was the Daytona road course this year. We'd never run on it before, never turned one lap. We did some simulator work, me and some other guys, and the numbers came back that we'd be going 210mph on the start–finish line, a big braking area. And we just said, "It's too fast." We added in a chicane. We made a safety call. That's an example of what we'll do now – we'd rather change the track than risk having someone's brake rotors exploding at more than 200mph.' Formula One had also decided the days of dangerous circuits were over. Today, new circuits are designed with safety at the

Fortunately, you do have a device that has a similar effect when it comes to reducing neck hyperextension: an airbag.

forefront, while older circuits have been modified. Imola's Tamburello, the turn that killed Senna, no longer exists; today it's a chicane.

Across the world, motorsport had reached a reckoning. Lessons would no longer be ignored.

The test as to whether the changes would work came in Bahrain on 29 November 2020.

CHAPTER TEN

Twenty-Seven Seconds

Let's rewind. Back to the start of this book. Back to a moment when the world held its breath.

Romain Grosjean has 27 seconds to live.

It's Sunday, 29 November 2020. The Bahrain Grand Prix has just started, the roar and sparks of the hybrid engines, thunder and lightning in the Gulf night. Twenty of the most advanced cars on the planet jostle for position. Through turn three, Grosjean darts to the right to avoid a wall of slower opponents in his path. He arrows for clear air, attempting to thread his car at 150mph through an ever-narrowing gap between Haas team-mate, Kevin Magnussen, and the AlphaTauri of Daniil Kvyat.

It's a gap that doesn't exist.

Grosjean's rear right tyre clips Kvyat's front left, sending him skidding off track in a shower of burnt rubber. He smashes into the safety barrier.

Freeze frame.

While the accident shocks the world, lessons from a thousand incidents before it have prepared for this moment. And while not everything has gone right, Grosjean has a fighting chance at life. As former FIA Foundation director general David Ward later said: 'From the moment of that collision, Grosjean began a journey through layers of defence the FIA has systematically developed to mitigate the risk of injury.'

These layers matter. Today, whenever there is an accident, the FIA investigates in forensic detail. 'I always talk about the Swiss cheese model of risks and losses,' Formula One expert

Mark Gallagher says. 'You have layers of regulations and compliance. First, the technology present. That's your Kevlar, crash structures and the rest of it. Then you've got systems and processes, such as safety cars, yellow flags. And, finally, in your cars, the last line of defence is people and their behaviours.' Imagine these layers as a block of Swiss cheese. All of the holes need to be aligned for disaster to strike.

All three layers – technology, processes, human decisions – will play a part in Bahrain.

★ ★ ★

Romain Grosjean's life is saved three times in the impact alone. As we know, energy doesn't just vanish, it has to be transferred somewhere. Grosjean's car hits the barrier travelling at 119mph, colliding at a 29-degree angle and 22-degree yaw. As it strikes, the nose of the car begins to deform and crumple. This is by design. At the front of the car, the framework is smaller, which allows it to deform more easily than the main protection for the driver, the near-impregnable 6mm carbon fibre and Kevlar survival cell. Crumpling the frame requires energy, repurposing the kinetic force before it reaches the survival cell and the driver inside. As Grosjean's 746kg Haas hits the barrier, the side impact protection bars also folded in on themselves, absorbing the shock.

This crumpling structure has constantly evolved. In 2007, Robert Kubica suffered a major accident at the Canadian Grand Prix, which prompted additional research from the FIA into impacts – not just in direct contact, but at oblique angles. They discovered the chassis tubes could actually break off in such accidents, prompting teams to submit various designs to solve the problem. Eventually, Marussia's design was chosen: a structure of tubes made from a high-performance carbon fibre that doesn't shatter on impact. The new structures, fitted to the cars in 2014, are able to absorb nearly 40kJ of energy.

While a Formula One survival cell is more advanced than would be seen in a typical road car, it doesn't mean the technology can't be used elsewhere. In 2019, 7,773 newborn babies in the UK needed immediate transfer to a specialist centre for urgent medical care. Such a transfer is incredibly difficult. Babies – particularly pre-term babies weighing only a few pounds – are a fragile and incredibly precious cargo that require a regulated environment with easy access to doctors and nurses. Unfortunately, a typical baby's incubator weighs about 120kg.

This is where an F1 survival cell comes in. Since the early 2000s, specialist 'baby pods' have been used to transport patients. In 2017, after a two-year partnership, Williams Advanced Engineering and Advanced Healthcare Technology launched the latest version of these devices, the Baby Pod 20. Using the same principles as a survival cell, the pod is made from carbon fibre and is capable of withstanding a 20*g* crash, with its passenger secure inside thanks to straps made from high tensile webbing. The baby is also surrounded by shock-absorbent foam, using a vacuum mattress that can be secured around the child. The pod has a CO_2 scrubbing system, and a clear, sliding lid to give medics a good view of, and access to, the patient, even if being transported by air. Better still, not only is the pod cheaper than an incubator, it only weighs 7kg. Today, the principles that protect F1's stars also protect children on their way to Great Ormond Street Hospital.

In Bahrain, the force of the impact reaches Grosjean's survival cell. It cracks in a few places, enough to trap his left foot, but ultimately it holds. Without the cell, he would have been dead in less than a tenth of a second. Now a second piece of technology steps in to keep him alive: the HANS device.

From onboard sensors (including accelerometers inside Grosjean's earpiece, moulded to his ear canal), we know the *g* force experienced during the crash was around 67*g*. For reference, stopping a car normally at 60mph is about 0.67*g*, a rollercoaster maxes out at about 3*g*, while NASA's centrifuge at the Ames Research Center only goes up to 20*g*. The highest *g* force someone survived came in 2003, when Indy 500 winner Kenny Bräck's car recorded 214*g* during a crash at Texas Motor Speedway. Bräck broke his sternum, femur, a vertebra and his ankles. It took him 18 months to recover.

Alone, the deceleration Grosjean has experienced would have been more than enough to break his neck. Fortunately, with his HANS device correctly installed, and his six-point seat belt in place to hold him into his custom-moulded seat, Grosjean is largely unharmed. The injury that killed Ratzenberger, Earnhardt and so many others is avoided. Directly behind the seat, another piece of design also helps reduce any risk of injury: the headrest. Far from the simple cushioned back of a passenger car, an F1 headrest envelops the driver in a U-shaped section of foam. Introduced in 1996, according to Sid Watkins, the device probably saved the life of Jos Verstappen the year it was instigated, when the Dutchman was sent spinning in a huge accident at Spa. In Bahrain, the headrest works in tandem with the HANS device to prevent Grosjean's neck from hyperextending.

The crush zones have worked. The survival cell has mostly held. The seat belt, HANS device and headrest have cushioned Grosjean as much as possible. Even so, his life is still in immediate peril. As Grosjean's car slices into the bottom two rows of the guardrail barrier, the upper section holds firm, creating a metal wall heading straight for Grosjean's crash helmet. Fortunately, the third life-saving intervention is ready: the halo device.

Brought in during the 2018 season, the halo is a simple piece of engineering: a 9kg curved bar of solid titanium, placed around the exposed head of the driver and connected to the chassis at three points. The simplicity of the halo belies its function. As mentioned earlier, it has the strength to withstand the weight of multiple cars piled on top of it. In tests, it's shown to be capable of deflecting a wheel fired at 150mph directly at the cockpit. Using simulation data of 40 real incidents, the FIA has determined the system increases survival rates by 17 per cent.

Despite the halo's clear advantage in saving lives, the design faced immediate criticism from drivers and spectators when it was proposed. Its detractors complained that it somehow took away the 'essence of racing'. Perhaps more importantly, when rescue tests were done at Silverstone, initially the halo proved as much a danger as a life-saver. 'We did cutting exercises with hydraulic cutters,' one of the rescue team told me. 'And because they were made of titanium, because of the way it cuts and because it was under tension, you literally got a razor-sharp metal boomerang flying off the car. We had to look at new ways of cutting it. So, sometimes the pushback is also about safety.'

These problems were overcome by adding a little carbon into the compound and working out different ways to cut the titanium. And ultimately, as Sir Jackie Stewart observed at the time, 'Preventative medicine is considerably better than corrective medicine. Mike Spence driving my car at Indianapolis in 1968 had a wheel come back and hit him [fatally] in the head. A halo would have prevented that, almost certainly.'

Within a year of its introduction, the Halo had proved itself. First in Formula 2, when Nirei Fukuzumi locked wheels with Tadasuke Makino. Fukuzumi's car cartwheeled up, and his right rear tyre glanced across Makino's cockpit, denting the halo. Makino, however, was unharmed. Then, at the Formula One Belgian Grand Prix, a pile-up on the first turn caused Fernando Alonso's McLaren to lift up, bounce on top of Charles Leclerc's halo and land hard on the other side. Once again, both drivers were unharmed. There is little doubt that, had the halo not been present, Makino and Leclerc would have been crushed.

In the Grosjean incident, all arguments against the halo vanish. The titanium wall refuses to buckle, forcing its way through the barrier and protecting the Frenchman's head. Grosjean himself had been a major detractor of the halo when it was introduced. The accident made a believer of him. 'I wasn't for the halo some years ago,' he later said from hospital. 'But [now] I think it's the greatest thing that we've brought into Formula One and, without it, I wouldn't be able to speak to you today.'

The car has done its part in saving Grosjean's life. But what about the object he crashed into?

* * *

Track design is an art form. Whether it's negotiating a tight city circuit or on a purpose-built track, everything is considered to try and make the route as safe as possible. The FIA has a series of grades, from 1 to 6, depending on the suitability of the track for racing. To even think about hosting a Formula One race, you need to be Grade 1: a designation held by 43 circuits in 30 countries.* This is the highest standard possible; just a brief glimpse on the FIA's website will give you a list of regulations for track layout, signals, floodlights, medical centres and, of course, barriers.

The point of a barrier is not to stop the car as quickly as possible; it's to make it take longer to decelerate. This allows the kinetic energy to disperse, meaning it's not all expended in one big hit. Barriers aren't placed by eye or scattered arbitrarily on a track. There are simulations – countless simulations – to work out what would happen if a car goes off at any point. There are standards for barrier placement, too, and inspections to check they've been done correctly.

You might think the optimum solution for all tracks would be to have a large run-off space, with gravel laid to slow down the car's motion and gently bring it to a halt. But science disagrees: for high-speed corners, you don't want gravel at all, as this would kick up a potentially deadly shower of debris. And while run-off areas are useful when a car is already turning, on straights you want the barrier *close* to the track instead. This means the car has less space to twist into the wall, preventing a head-on collision. Rather than having a softer barrier, on straights the designers will use a hardy, resistant material like concrete, which can deflect the car

* In case you're wondering, the Circuit Ricardo Tormo in Valencia is Grade 2; as is Circuit de la Sarthe in Le Mans. The Bahrain International Circuit is, obviously, Grade 1.

along its side and convert that kinetic energy to heat – through friction. While concrete can be deadly head-on, when used in this way it's a relatively safe and effective method of bringing a car to a halt.

These are far from the only barriers that have been used in motorsport. The earliest were straw bales – a pretty cheap and easy way to reduce momentum. Unfortunately, these could catch the car and cause it to flip or spin, resulting in huge whiplash injuries for the driver. They're also flammable. In 1967, these deadly properties combined at the Monaco Grand Prix, when the Ferrari of Lorenzo Bandini flipped after clipping the straw bales lining the harbour. The impact ruptured the fuel tank and the bale ignited, leaving Bandini trapped upside down amid burning straw. He died three days later with 70 per cent of his body covered in third-degree burns. Formula One banned straw bales from ever being used again.

Next came wire fencing, which could deform to absorb an impact. But this wasn't a huge improvement: often the fence would wrap around the car, trapping the driver in a deadly net and blocking the marshals from coming to his aid. At the 1981 South African Grand Prix, Carlos Reutemann was almost strangled to death by the very fences put there to protect him. Today, wire is only used as netting to catch flying debris. Instead, you're more likely to see barriers of tyres bolted together, basically creating a bouncy wall that dissipates energy. There are also purpose-built Tecpro barriers that work in a similar way, with the red-coloured pieces deforming and the grey having a little more reinforcement to catch the car as it leaves the track.

The most common sight on track at low speed, low-risk sections is the type that Grosjean hit: a guard rail, also known as an Armco barrier – a name that comes from 1899, when The American Rolling Mill Company was founded. An Armco is a long, seamless beam of steel, identical to the type you'll find along the edges or central reservations of most major roads. If you look closely you'll see the metal isn't flat: it's designed in a 'W' shape. Thanks to mathematics, it's very difficult for a material to bend in two ways (think about holding a slice of pizza – you bend it in a curve, making it less likely to sag). This allows the beam of the barrier to absorb energy, with the supports and posts ready to give way and help take some of the force from the impact, too.

There's one last type of barrier that needs to be mentioned. Even before the death of Dale Earnhardt, barriers had fallen under the spotlight in the US. As the tracks are high-speed ovals, most were lined with concrete, but this wasn't enough to offer full protection and alternatives were trialled. In the 1990s, IndyCar had started to look at barriers; by 1998, 41 drivers had died at the Indy 500 and its owners wanted to do something. The sport had started using the Polyethylene Energy Dissipation System (PEDS), which would deform whenever it took a hit. Unfortunately, the PEDS barrier could do the same thing as straw bales: catch the car and cause it to pivot or, occasionally, lift it in the air. More work still needed to be done.

The man who came up with a solution was Dean Sicking at the University of Nebraska–Lincoln. Instead of plastic, Sicking advocated steel. Initially there was hesitation from both NASCAR and IndyCar – after all, everyone knows steel is harder than plastic – but eventually Sicking persuaded both competitions to try his new design. It was called the Steel and Foam Energy Reduction (SAFER) barrier. As the name suggests, SAFER involves square steel tubes, which are welded to form a flush wall with foam absorbers behind to help absorb impact. These barriers had two major advantages: they could take a hit from the car without causing it to bounce back into traffic, and they reduced the chances of a car pivoting or being sent flying into the stands. Tests show SAFER barriers are capable of extending a collision's time from a tenth to two-tenths of a second – reducing the force of impact felt by the driver by up to 80 per cent.

After the death of Earnhardt, the SAFER roll-out was accelerated, with the first barriers installed at Indianapolis in 2002. Ten days later, at an IndyCar race, they were 'tested' for the first time when Robby McGehee ploughed into them. McGehee had become an inadvertent guinea pig, allowing his crash to be directly compared with another, by Eliseo Salazar, shortly before the barrier was installed. The acceleration experienced by McGehee was 60 per cent less than that suffered by the unfortunate Salazar, whose injuries were so severe he suffered a torn vertebral artery that forced his retirement from IndyCar. By 2005, all NASCAR tracks were using SAFER. 'The SAFER barriers were a huge thing,' Jacuzzi says. 'We went from having concrete on the walls to having them almost everywhere.'

Today, the principles behind Sicking's SAFER barriers are used around the world and 27 US states have adopted his design for their roads. In 2005, President George W. Bush presented him with the National Medal of Technology and Innovation, the highest award for a US inventor. In addition to saving hundreds of thousands of lives with his barriers, Sicking has also developed barriers for American football fields and ice rinks. Whether it's Formula One or NASCAR, the barriers designed to stop high-speed cars have a direct link to the barriers that protect us on the road and beyond.

★ ★ ★

Back in Bahrain.

Less than one second has passed. Grosjean's car is still ripping through the barrier. At the rear of the Haas, the tangle of twisted metal creates jagged edges that slice his car in half. In a thousandth of a second, hig-octane fuel spills, ignites and explodes in a spectacular fireball. The car vanishes from sight, lost in a sheet of burning death. Romain Grosjean is trapped in a broken wreck at the heart of the inferno.

'I didn't lose consciousness,' Grosjean later told French TV channel TF1. 'To get out of the chassis, I was able to undo my belts. The steering wheel was no longer there, it probably came off in the impact … I saw my visor turning orange, I saw flames around me. I thought about a lot of things, including Niki Lauda, and I thought I didn't want to end up like that. I couldn't finish my story in Formula One like that.'

The fire that surrounds Grosjean is terrifying – but it could have been so much worse. Later, experts looking at the fire would conclude it wasn't anywhere near big enough to represent the full 100kg of fuel that would have been in Grosjean's car. As mentioned earlier, Formula One cars don't have a normal petrol tank but, rather, a Kevlar re-enforced fuel cell. 'With fuel cells so incredibly strong,' F1 sporting director Ross Brawn was later quoted in *Autoweek*, 'I suspect that [the fire] came from a ruptured connection.' The official report would later prove Brawn's assessment correct. The power train assembly and survival cell had separated, dislodging the fuel tank inspection hatch and ripping the engine fuel supply connection away. The fire came from these two points.

There is a fire suppression system in Grosjean's cockpit but that isn't enough to deal with the blaze around him. Instead, his main protection comes from his clothes. In addition to his fire-resistant helmet, he is dressed in a race suit made from Nomex, a synthetic polymer released by DuPont in the early 1960s.

On a chemical level, the suit's fibres consist of long chains of aromatic polyamides, or aramids. This chain is linked by amide groups (CO–NH), which provide a strong covalent bond. This is the same way natural materials, such as wool and silk, or synthetic materials, such as nylon, are put together. But aramids also have groups of carbon atoms arranged in a hexagon called an aromatic ring. This provides a lot of resistance to the environment.

Aramids really took off in 1964, thanks to chemist Stephanie Kwolek, who was only working at DuPont's lab to try and earn enough money to put herself through medical school. Kwolek had been tasked with trying to find an alternative to steel fibres in tyres but, rather than making the clear, viscous liquid the company was looking for, hers was cloudy, flowed easily and was 'buttermilk in appearance'. Usually, such bad batches were thrown out, but Kwolek convinced the technician to spin it into fibres anyway. It was the discovery of Kevlar, a substance five times stronger than steel.

There are two types of aramid, depending on how their structures are linked together. Para-aramids, such as Kevlar, are incredibly strong. Nomex, however, is a meta-aramid. These lack a para-aramid's strength but have high melting points (about 370°C), and are hard to set on fire – making them perfect for fire-resistant suits. Today, you'll find Nomex worn by racing car drivers, firefighters and even astronauts. While it won't resist really high temperatures forever, the suits are tested at temperatures in excess of 600°C to make sure they give adequate protection for long enough to escape. When CNN asked DuPont for a demonstration, they were taken to the company's Thermo-Man, a mannequin equipped with 122 heat sensors that can be repeatedly subjected to a huge fireball. When the mannequin wore cotton, it had burns over 74 per cent of its body with a survivability estimate of 35 per cent. After the mannequin was dressed in a Nomex suit, only 34 per cent of its body experienced burns with a survivability of more than 90 per cent.

Nomex is impressive; but the suit still only buys Grosjean a few extra seconds of life. At this point, technology has done all it can. Now it's time for processes to come into play.

Around 5.5 seconds after the incident, the race is red flagged. Thanks to the work of Watkins and others, aid is there immediately. On the first lap of the race, the pack is always followed by a high-speed medical car. In Bahrain, it is driven by former British F3 champion Alan van der Merwe with Dr Ian Roberts in the passenger seat and another track medic in the rear. In eleven seconds, van der Merwe pulls the car alongside the scene, and Roberts is out of the cab and rushing towards the fire. van der Merwe, meanwhile, makes for the boot to grab an extinguisher. A local doctor, climbing out from the rear of the car, prepares the trauma bag. Already, a fire marshal from the Bahrain Civil Defence, Sergeant Joby Mathew, has arrived on the other side of the barrier. A second, Corporal Thayer Ali Taher, is rushing across the track to Roberts' position, hauling a second extinguisher.* None of the men hesitate for a moment.

---

* The two marshals both received promotions the next day; Taher to sergeant and Mathew to sergeant major.

Mathew opens his extinguisher. There is no way he will be enough to put out the fire … but he might be able to push the flames away and give Grosjean a chance of escape.

★ ★ ★

John Trigell is well versed in motorsport safety. In the early 2000s, he joined McLaren and became safety manager for its road cars. Eventually, he'd spend 15 years in Formula One, including at Renault during its glory years with Fernando Alonso's back-to-back championships. While there, he formed a safety group for all F1 teams as, despite the efforts of Watkins on track, the factories were lagging behind. 'To give you an idea,' Trigell recalls, 'half of the UK teams didn't have a dedicated safety person, and they were running 700 people in their factories. Before that group, they were sending people like HR managers to take responsibility for safety.' Perhaps there's no better illustration of how little heed was paid to health and safety than Trigell's own background. Prior to McLaren, he'd worked for 20 years as a cabinet maker at a furniture company and was initially hired as a logistics expert. He took over the health and safety role at McLaren – and completed his certificates to do it – because the team literally had no one else.

Although the death of Senna had focused F1's mind on car safety, it hadn't had a similar effect on team wellbeing. 'Back then, by the time a member of the pit crew hit thirty-five to forty years old, their wrists were shot because of the wheel guns, which are extremely powerful, and their knees were gone too,' Trigell says. 'So, we had to implement an awful lot of measurements of vibration, then went to the constructors and got them to use gloves for their pit crew that could absorb those shocks, and gel pads for their knees. Gun time was cut down. These days, if you watch a pit stop, there's a lot of smaller wires that attach to the guns. These measure performance of the person – how fast they are making a pit stop – but also the speed of the guns. It's to stop people wrenching their wrists.'

From basic team wellbeing, Trigell moved up to overall safety in a race. He's now director of operations for MDD Motorsport Medical, which is responsible for medical services to a range of championships, including Formula E. It's his job to recruit doctors to look after racers, much like Watkins and Roberts. 'There isn't a specific qualification that's used for race medics,' Trigell explains. 'We use quite a few different doctors. We've even used a special ops doctor who was a field surgeon, training the FBI and CIA. We used him as a paddock doctor and then got him assessed by the FIA. The doctors we recruit all know their stuff. They just have to be able to do it on someone in a car.'

The idea of a single 'race doctor' is a misnomer, Trigell explains. 'We take sixteen doctors to a Formula E race. This includes the paddock as well as the track. We've got an advisory doctor who's on track as part of the extraction crew, a local on-track doctor and then a chief medical officer for the FIA, who sits in race control. So, you have two on-track doctors, an incident commander and four extraction people standing by.' This is a little different from Formula One, but the emergency process is still the same. 'The accelerometers on the car will alert the crew there's been a high *g* impact. You head out, and you address the driver. If there's fire, it'll be a rapid extraction. We can get a driver out in thirty seconds. This is always the priority for a medical crew [after their own safety]: to get the driver out as soon as possible and get them treated.'

The medical drill is practised before every race. 'No one in the world has the experience a race doctor has,' Trigell says. 'We test our people an awful lot. To maintain being the best, you have to practise, practise and practise. At every single event, they practise. At Formula E, we practise a full red car [electrically live],* on track, with ambulances and one of the drivers. The FIA don't tell us what the type of incident will be. They dictate whether we're, for instance, practising if the car's gone up a wall. We'll also get a shout-out that someone has collapsed in a garage, or there's an injury in the pit lane: it'll be a practice drill, but we won't know. We measure our response times and extractions. And, of course, there's a second local team on standby, because if you've got multiple incidents on track, then you need a second rescue team. In the early days of Formula E, we had bad experiences in some countries. But we assess the teams and if they don't pass muster we'll step in. They won't be the second team any more.'

Around the pits, you'll also find that everyone knows how to save a life. 'At a Formula E event, you'll find something in the region of twenty-five to thirty defibrillators,' Trigell says. 'Every single person in each garage is trained to use them. They'll all have electric safety kits and know how to use them, and they're all licensed. Anyone working on or around that car has gone through training on what to do. It's that strict.'

---

* Formula E extraction crews carry a custom-designed rubber drape, moulded into the shape of the car, that can be deployed instantly to nullify any electrical danger. Fortunately, it has never been used in a real incident.

Let's take a look inside a medical car. In the driver's seat will be a professional race driver, surrounded by an array of screens giving constant updates of the race situation and drivers' track positions. They also get live biometrics from the drivers' gloves and accelerometers, and have an intercom and hand-held radios to ensure constant communication with race control and each other. In Formula One, Roberts gives a running commentary during every race, relaying what's going on ahead so race control knows exactly what's happening from his perspective.

The medical kit available is pretty standard, the type of equipment found in any ambulance or trauma centre in the world. In Formula One, the medical car's boot contains two kit bags: one containing the oxygen and airway kit, and one a major trauma kit – both different colours for ease of identification – along with electrical safety gloves, burns dressings and two fire extinguishers. Roberts also has a smaller bag with him, containing everything he might need immediately: intubation to secure the airway and tourniquets to staunch bleeding. 'The medical car also has a supply of proper sedatives,' Trigell says. 'If a patient is seriously injured, they'll need to be sedated before they can be flown to hospital by helicopter.'

All of this comes together in Bahrain. Before the race even began, Roberts, van der Merwe, Taher and Mathew all know their roles. They've talked about what to do if a car goes through a barrier. If a car is on fire. If someone's life is on the line. It has become a well-rehearsed drill, a choreographed dance routine. Van der Merwe has never had to use the medical car's extinguisher in his 11 years of service, but has talked about it so often in practices and briefings that his rush for it is instinctive. The reason the car is with Grosjean so quickly is thanks to preparation, too: van der Merwe had taken a pre-agreed shortcut to avoid the track's turn one and keep up with the cars.

This training does not diminish medical teams' astonishing personal bravery. It takes courage to run towards hell. It takes presence of mind to know what type of extinguisher is needed and how to tackle a fire effectively. And it takes years of fostered trust and teamwork to know, without doubt, that the man next to you is doing all he can.

Eleven seconds after the accident. As the medical car arrives, Mathew begins to tackle the blaze from his side of the barrier.

Nineteen seconds. Taher finishes his sprint towards Roberts, having hauled the hefty extinguisher across the track. Roberts directs him to blast where he sees something in the flames. 'The guy that was with me,' Roberts later told *F1TV*, referring to Taher, 'he extremely bravely crossed the track and the blast was directed exactly at where Grosjean was.'

Another medic has rushed on to the track. At 25 seconds, another three marshals have entered the picture from the helicopter circling overhead.

Roberts sees something, a shape in the blazing wall in front of his eyes. He advances towards the roaring furnace, ignoring the intense heat licking his uncovered face. Through the smoke and flame, past the white haze from Taher's extinguisher, something moves. Roberts reaches out towards the barrier …

★ ★ ★

'I try to go up a bit more on the right, it doesn't work,' Grosjean later told Formula One's website. 'Go on the left, doesn't work so I sit back

down … It couldn't finish like this, no way, so I try again. Then there's the less pleasant moment where my body starts to relax, I'm in peace with myself and I'm going to die … then I think about my kids. And I say, "No, they cannot lose their dad today."'

Formula One drivers aren't just allowed to hop in a car and go. As part of pre-season testing, they have to prove they can get out of the car in a hurry. Before the halo, the time limit for escape was set at five seconds and often resulted in a damaged wing mirror or barge board in the hasty scramble. Since 2018, the time limit is seven seconds, as the halo device means it takes slightly longer. (Although, given it stops you being decapitated, it's hard to argue with an extra two seconds' scramble.)

Grosjean pulls his foot. His shoe comes off, stuck somewhere in the survival cell's depths, but he doesn't care. He reaches up, hands in pain – his gloves are the weak point in his flame protection and his hands are badly burned – and grasps the side of the car, forcing himself out of the seat. Amid the blinding glow, he can see his gloves are beginning to melt. He jumps from the cockpit, finds himself on the barrier and feels Roberts' grip pulling him to safety …

Twenty-seven seconds after hitting the barrier, Romain Grosjean escapes.

If the technology in his car had failed, he would be dead. If the medical car and marshals had not been there, he would be dead. If he had not had the presence of mind that comes through practice and training, he would be dead.

But he is alive, alive, alive.

The danger has passed, but Grosjean's night has only just begun. Taking his first steps towards safety, he is convinced he must be a running fireball. He's not – the Nomex has protected him. Roberts tugs at his overall, letting Grosjean know he's with friends. The driver pulls off his gloves, convinced that his skin must be a bubbling, melting ooze. It isn't. It's red, swelling and painful, but will recover in time.

By now, van der Merwe is advancing with the extinguisher. He and Taher blast their jets at Grosjean and Roberts, ensuring that neither are smouldering from their close encounter. The doctor and patient enter the medical car. A cold compress goes on Grosjean's hands. His shoeless left foot begins to ache, the body's fight-or-flight response settling down and giving way to an agonising pain.

'The ambulance is coming,' Roberts tells Grosjean. 'They're going to come with the bed and you're going to be OK.' Already, it's pulling up on the other side of the track.

'No,' Grosjean says. The defiance of a man who has just seen death. 'Now, we walk to the ambulance.'

'No, no, the bed is coming … '

'No. No. No! I walk out of the car! We are *walking*.'

And, with the aid of his heroes, Grosjean walks. To the ambulance. To the helicopter. To the Bahrain Defence Hospital. To the rest of his life.*

* * *

There's an African proverb: it takes a village to raise a child. It takes a paddock to save a driver.

'The things that happened [in Bahrain] are the accumulation of all that had gone on before,' Roberts later said of the

* This chapter was based on camera footage, expert opinion and numerous in-depth analyses from motorsports journalists, and subsequently confirmed by the FIA report into the accident, published in March 2021.

Grosjean accident. 'There are many people in the background who worked extremely hard to get things to where they are now. But no one is sitting back. It's constantly evolving. Sid [Watkins], Charlie [Whiting, race director and safety delegate], Andy Mellor, an engineer who's done loads of safety work, Laurent Mekies [sporting director of Alpine F1] who was instrumental in the halo development. All of these guys were there. We just did that extra bit that brought their work together.'

At the time of writing, Grosjean's rehabilitation continues. He is expected to make a full recovery and there are even plans for him to race in IndyCar next season. After the crash, he was discharged from hospital after three days. His foot was injured and his hands were swathed in bandages for weeks. He would continue to use painkillers for months, and would struggle to move his hands and fingers into 2021. Psychologically, the accident stayed with him, too; two weeks after the crash, he started hearing the sound of fire again and smelled the burning of carbon fibre. Many of his memories of the incident are repressed. 'I told everyone there was no fire in the cockpit,' he later said. 'Then I saw the images from the onboard camera, which have not been made public yet. The fire [was] everywhere. Just everywhere.'

This brings us to the final remarkable feature about Grosjean's accident: how we know exactly what happened.

When Senna died, there was only a single camera feed of the accident, taken from a distance. When Grosjean lived, the incident could be reviewed from cameras placed on every car and at numerous angles on track, giving investigators a high-definition, 400-frames-per-second overview of the events that unfolded.

Since 1997, all Formula One cars have contained an accident data recorder – essentially a black box to record how safety equipment has functioned. As each team has identical liveries, one car has a black camera, the other fluorescent yellow so they can be distinguished. In 2016, they also introduced a camera that faces the driver, which shows exactly what decisions were made so lessons can be learned. But, importantly, this information isn't just stored on the car; it's also transmitted live to race control.

And this ability, to send data quickly and accurately, has directly led to another technology set to revolutionise our lives. Without the lessons learned after the death of Ayrton Senna, or the lessons that went on to save Romain Grosjean, we wouldn't have the latest generation of self-driving cars.

CHAPTER ELEVEN

Rise of the Robots

Every year, Lucas di Grassi holds an unusual race against one of the best drivers in the world. The Formula E world champion and former F1 driver lines up on track, races off the line and puts in the fastest time he can. Then his rival completes a lap, aiming to beat di Grassi's marker.

The twist is that no one is in the other car. Di Grassi is racing against an artificial intelligence (AI).

So far, di Grassi is undefeated. But the AI is catching up. 'Today, an autonomous car is only three per cent slower than me,' he says. 'It's a bet that I'm going to lose. When we started racing AI in 2018, they were twenty per cent slower per lap. I'm looking forward to when an AI can beat a professional racing driver. And it's only a matter of time.'*

Di Grassi's latest rivalry stems from Roborace, a competition made entirely of autonomous, or self-driving, electrically powered vehicles. From 2017 to 2019, he was the championship's CEO, working closely with Formula E to test the technology. 'I'm a racing driver,' he says. 'My dream was always to drive fast cars. But I saw that motorsport was a research-and-development platform. And nothing comes to my mind more than helping to have driverless, electric vehicles going cheap and safe around cities. Roborace is to have a top-down view, to find that algorithm that is faster than any driver, then scale that.'

---

* Di Grassi was speaking at a webinar hosted by All About Circuits.

Autonomous vehicles came to prominence in 2004, when the US Defense Advanced Research Projects Agency (DARPA) initiated its Grand Challenge. While this was to develop autonomous vehicles for military use, it ignited a much wider interest in the concept. Fortunately, this happened just as computer processors became small and powerful enough to run vehicles. In the past decade alone, governments and private companies have invested $100 billion in getting AI-controlled vehicles to work.

The driver for this technology, di Grassi explains, is clear: when you remove a human, you save money. 'There are two aspects. When I take a taxi, about half the cost is the driver. When you take the driver out of the equation it becomes cheap. That's why there's a race for automation: the first company to do it gets the data, exponential growth and dominates the market. The other aspect is that when you buy a car, 95 per cent of the time it sits still. It's a depreciating asset that takes up real estate. It's not going to be like this in the future. You'll request an autonomous car and it will pick you up to take you somewhere. Then, someone else requests it.' In effect, it would be Uber without the driver.

Currently, most autonomous vehicles are too slow for demanding commuters. Sport, however, allows you to flip the equation: starting with technology that's faster than needed and slowing it down to a road-safe level. 'That's why Roborace is so important,' di Grassi says. 'The first steps of automation, in airports, buses and existing routes, are already here but they've started from the bottom up. We're starting at the most difficult, then bringing the technology down to commercial vehicles.'

Roborace initially used a custom vehicle, Robocar, for public demonstrations. It looked awesome, although that's to be expected: it was created by movie designer Daniel Simon, the mind who envisaged the light cycles from *Tron: Legacy*. Half Batmobile, half fidget spinner, Robocar was packed with clever little touches. Rather than have a single engine, the Robocar had four independent 300kW motors and a 540kW battery, all run by an Nvidia PX2 brain capable of 24 trillion operations a second. These motors allowed torque vectoring – control of how much force, in forwards or reverse, is given to each wheel at any time – which brought a whole new level of capability when making a turn. The car was pretty good in a straight line, too: in 2019, it set a two-way speed of 175.49mph, making it the fastest autonomous car in history.

But while flashy looks are good for grabbing headlines, the real purpose of Roborace is as a development platform – a system to test AI software. Now in 'season beta', the teams race using another custom-built machine, the DevBot 2.0, which can be driven by a human (such as di Grassi) as well as by an AI. The racers – including teams from independent companies as well as the likes of the University of Pisa and MIT – all use the same hardware and car design. This means any difference is purely down to the AI software the teams install. You have to out-nerd your rivals to claim victory.

Yet racing isn't just the perfect testbed for autonomous technology, either; it's an originator of it. Without motorsport, AI cars wouldn't be able to fully visualise our world.

★ ★ ★

Bryn Balcombe caught the racing bug early. Growing up in Biggin Hill, where south-east London gives way to Kent, he spent the summer before he went to university working at the old 'west camp' portion of the local airfield. Once an aerodrome used during the Battle of Britain, it was now home to Bernie Ecclestone's planes and Balcombe found himself helping with their onboard cameras. 'At the end of the summer, they took me out to the Portuguese Grand Prix,' he recalls. 'It meant I turned up late to my first day of university but I thought it was worth the sacrifice.'

Balcombe studied mechanical engineering and vehicle design at the University of Hertfordshire, going on to specialise in road safety. During his time there, he took a year out to work on airbag design for Lotus Engineering. Once he graduated, his thoughts soon drifted to motorsport. 'It was around 1996,' he recalls. 'Formula One was looking at what it could do to improve safety post-Senna. So, I reached out to friends and looked at how I could combine my two passions.'

Balcombe began to work on onboard camera communications. At the time, Formula One cars had no direct car-to-ground link: everything was relayed through a helicopter. This limited what could be transmitted from a car. So, while there were onboard cameras with Senna when he crashed, the footage from them doesn't exist: the race's TV director had to choose which action to show and, at the time, had switched to another view. While the camera wouldn't have saved Senna's life, its footage would have been invaluable to accident investigators.

'My target was to say, "How can we do this, just using ground-based antennas?"' Balcombe says. 'Cellular mobile communications, mobile phones, were just coming out at the time, but they weren't designed to communicate with objects at the speeds the cars were moving, or to function in the high multi-path environments we were working in, with metal fences and barriers. We had to come up with an innovative solution.'

As mentioned previously, the steady stream of telemetry data from a racing car is astonishing, from encrypted RF transceivers and TV cameras to the sensors giving microsecond feedback. For safety purposes, even the position of the steering wheel, brake and accelerator pedals are sampled and stored. Yet the challenge is how to pass over this data. Think about trying to use a mobile phone in a car or train while travelling at a normal speed – sometimes it struggles. An F1 car is going at *three times* that speed and sending *100 times* that amount of data. This wouldn't be possible over today's existing mobile networks, never mind those of 25 years ago.

Formula One's solution to the problem is, in effect, to create its own mobile phone network before every race. An optical fibre ring is laid around the circuit and as the cars move they connect to a network of antennas. Each team has its own IP addresses and subnets, too, allowing the sport to handle the equivalent of an entire city block's worth of data coming from just 20 cars. The speed of these connections is astonishing. Since 2017, during practice sessions Mercedes has been measuring tyre performance using a fast wireless uplink that can transmit data as soon as the car enters the pitlane. Once the car gets within 4m of the garage, it downloads its information at 1.9 gigabits a second. Theoretically, this means it could stream four HD feature-length movies from the car to the garage in less than a minute. The team's flash array servers then allow them to rapidly access, analyse and encrypt data – something that could be vital for better communications and cybersecurity in aeroplanes, ships and hospitals.

Back on track, it's this technology that enables constant two-way communication with the driver. It also allows race control and teams to know the exact position of the cars relative to each other, both for safety and so they know when a car can activate DRS. And, of course, it lets the stewards (and the folks at home) watch the race from the driver's viewpoint in high definition. 'It's one of the only patents that Formula One management has ever filed,' Balcombe says. 'I'm quite proud to have my name on that. It's on its fourth generation now and it's the reason why you see such stunning onboard camera footage. But it's also why we know what happened with the Romain Grosjean collision – all of those cameras are available to record, transmit and give us an evidence base.'

This leads directly to safer autonomous driving. Such HD cameras provide the 'eyes' of an AI car – eyes that are better than a human's. Today, Balcombe is chief strategy officer for Roborace, as well as founder of ADA, a non-profit dedicated to using AI to end fatalities on the road. 'If you look at the Grosjean accident, it's quite clear that Romain didn't see the car behind him. And therefore, when he pulled over, he ended up in a collision.' This is where AI has the advantage, Balcombe explains. While humans have two eyes pointing in roughly the same direction, an AI vehicle can have multiple cameras to track objects around it, as well as a host of other devices to tell it about its surroundings. Robocar, for example, has six AI cameras, GNSS positioning,* five Lidar sensors (which bounce

* Global Navigational Satellite System, the generic term for satellite navigation. GPS is just a brand name for the version operated by the United States Space Force, which is a very real thing even if it sounds pretty goofy.

lasers off objects to determine their range) and 18 ultrasonic sensors. This means the car can determine its distance to any object, in any direction, with the car knowing where it is in the world to within 1cm. 'What's really fascinating for us is that when you bring AI into motorsport, you have the capability to see in three-hundred-and-sixty degrees,' Balcombe adds. 'The kind of accident Grosjean had would be considered unacceptable in autonomous vehicles because they have no visibility limitations.'

And visibility is crucial to ensuring an autonomous car is safe, Balcombe explains. 'Autonomous driving tasks can be sorted into perception and planning. And the perception part is really why there's been a gold rush recently for autonomous cars. We have deep learning algorithms that can look at a camera-based image and classify objects accurately. The autonomous driving industry looked at that and went, "Well, video is just a sequence of static images." If you can accurately detect humans and cars, you should be able to track them, too. That gives us the ability to drive autonomously.'

Much of the work on AI in the past 10 years has, therefore, focused on building up a better picture of what's happening around the car. This is called a 'local world model', and enables the software to assign an ID to an object, classify it (human, lamppost, car, *etc.*) and predict how it's going to move. 'Once you have the local world model where objects are located, the next task is planning,' Balcombe says. 'You know, what's the action you need to take in that environment, what's the appropriate level of risk? It's incredibly challenging, though, because not only do you need to detect the object, you need to predict its future motion, quite a few seconds in advance. That's still a very difficult task for autonomous driving.'

The way to develop this skill is to provide a structured environment with defined parameters. That's exactly what a racetrack provides. Each Roborace event features two types of race. In the first, it's a simple dash around the track – best time wins. In the second, the AI has to navigate around a real-world track filled with virtual objects in augmented reality, some of which are power-ups, some of which are time penalties. If this sounds a little too close to a high-speed version of *Mario Kart*, it's all for a good cause: it allows the teams to push the boundaries of their AI and see what helps it make the best decisions.

Unlike other autonomous driving pioneers, such as Tesla or Google, the teams are happy to share the lessons they've learned with everyone. All you need to do to take part is download the base algorithm and the car's virtual twin, and see if you can come up with a better solution than someone else. 'There's no other environment at the moment where multiple AI systems are tested and developed while sharing the same environment,' Balcombe explains. 'And that's crazy if you think about it. Different brands of autonomous vehicles are all going to end up on the road together and they've never seen each other, never interacted with each other! But if we take a collaborative approach now, then you can share information between those vehicles.'

This sharing is imperative if autonomous cars are going to become a reality – and, once again, falls back on the tech Balcombe developed in the 1990s for Formula One. 'If you see AI in racing, safety is likely where it'll be used first,' he says. 'No one wants driver aids that are a detriment to racing from a sporting sense. But what you could see is AI introduced for safety: red flag situations, driving in the pitlane, or if there's a safety car. You could even use it in a race-control environment. In 2002, at the Circuit Paul Ricard, we developed a blueprint for all the safety systems that need to be deployed at future racetracks. That includes being able to have an automated camera system that follows each car, both for teams and for stewards. Twenty years later, it's easy to envisage how new safety benchmarks could be set using new trackside sensors and AI.'

'You could go further still,' Balcombe continues. 'You could have AI monitoring those video feeds, looking at behaviours, fulfilling the role of race stewards. And that's one area that we can really push: how can we use a race's infrastructure to improve safety?'

* * *

If there's one race that clearly illustrates the need for Balcombe's vision, it's the 2017 FIA GT World Cup. The event took place on the Guia street circuit, in the heart of the twisting, mazy roads of Macao, a former Portuguese colony turned gambler's den at the mouth of the Pearl River in China. Unlike most street circuits, the Guia course hasn't been modified since it was established in 1954 and is essentially a series of long straights and tight bends. There is nowhere to go if you make a mistake except the wall.

During the qualifying race, Daniel Juncadella crashed into the side barrier and stalled. Unfortunately, the accident had happened on a blind turn: none of the racers behind him could see the unavoidable roadblock that lay ahead. In a matter of seconds, a horrifying domino effect began as rest of the field ploughed into Juncadella and each other, leaving a solid mass of crunched steel, spinning wheels and dazed drivers. When the dust settled, 16 of the 20 cars on track had collided – including di Grassi, whose car had somehow ended up on the bonnet of a rival. Although no one was hurt, it took more than an hour to remove the carnage from the track. Several million dollars' worth of precision engineering had been totalled.

With connected AI, the accident wouldn't have happened at all.

'That pile-up continued because the drivers didn't know there was a car there in the first place,' Balcombe explains. 'Vehicle-to-vehicle communication creates an infrastructure that exceeds perception levels of a vehicle, even an AI one. If you can't see around a hairpin, you have no idea there are objects in place. But if you can extract and share that information from other cars, then you can make safe decisions based on it.'

This extends to the real world. If autonomous cars can talk to each other, they can make better decisions about how they drive on the road, minimising the risk of collisions and improving traffic flow. And it doesn't just impact safety, either: AI cars can help save the planet.

Autonomous vehicles have six levels. Level 0 is no automation at all, with the human in complete control. This builds, taking on tasks such as controlling speed or assisting with parking, up to Level 5: full autonomous driving, moving seamlessly between environments with different rules of the road. (Tesla, for example, currently operates at Level 2, while Roborace is Level 4). These levels leave a lot of room for humans to use AI assistance without fully giving over control.

At The Ohio State University's CAR, David Cooke is already exploring some of these options. 'We've been working on an $8 million programme sponsored by the US Department of Energy,' he says, 'Things like lane-keeping and adaptive cruise control are Level 1, headed to Level 2 technologies. But the US government said, "autonomy is coming and we're going to use it to reduce crashes and road congestion. That's wonderful. But can we also use it to save fuel?" So, they charged us with finding a way to use Level 1 and Level 2 autonomy to reduce fuel consumption by twenty per cent, and gave us three years to prove it.'

The CAR scientists decided that the best way to cut down on petrol was to look at how vehicles connect with each other and their environment. 'Once you use your satellite navigation, we know the route you're going to take, what the hills look like, where the traffic signals are. And we can programme the car to control your power train and optimise its operation.' Cooke gives a simple example: you can teach a computer that a red light means stop and a green light means go. 'If you connect to what's called the SPaT – Signal Phase and Timing – of the traffic light, your car can know when the lights will be red and slow down or speed up to avoid it. Imagine a world where traffic lights are connected to cars. If the car knows your route, you'd never stop at a red light again.'

A world without traffic signals is also good for the environment. 'Stopping at a red light and accelerating means a lot of energy is wasted,' Cooke says. 'The best thing you can do is cruise at a constant pace. There's no reason to drive at eighty miles per hour and end up in a traffic jam when you can drop to sixty and go right through a light without stopping.' Less energy wasted, less fuel burned.

And Cooke isn't finished. 'That's just traffic lights. Now, imagine you could model the entire world. There's all kinds of big data problems and machine-learning algorithms that come into play. A simple one to understand is hills. If you know when a hill is coming up in your route, you can use your battery to get up the hill and then go full regen on the way down. Just knowing when a hill is coming up can save two to six per cent on fuel economy. And that's huge! We've already demonstrated more than twenty per cent [fuel saving] across a wide variety of routes, from cities to highways. It looks like we're going to achieve about a twenty-five per cent increase in fuel economy. Not everything is ready to be deployed on cars yet, there's still a lot of work necessary. But they're on the cusp of being built into production vehicles.'

★ ★ ★

As with any developing technology, Roborace has had a few teething troubles. In 2017, during its first public competition in Buenos Aires, one of the cars misjudged a corner and crashed. Far from being a disaster, this just added to the evidence base about how to prevent future incidents. And, while the crash took the headlines, it overshadowed the fact that the other DevBot racer completed the course at speeds of up to 116mph. Similarly, in October 2020, a car from the Schaffhausen Institute of Technology decided to start a race by immediately turning right and driving itself into a wall.* Yet anyone who uses this as ammunition against AI cars is

* If you want the technical explanation of what happened, the car developed a 'not a number' fault – basically a computer being asked to do an impossible calculation – during the previous lap, which locked the steering control to the right.

missing the point. These are prototypes. This is *why* you test. This is how you learn.

'Even in the past four years when we've been running Roborace competitions, we've seen the technology advance massively,' Balcombe says. 'We've had three generations of computer platform in just four years. The turnover time for improvement and innovation is much faster than the automotive industry is used to. What's going to be really interesting is when those two worlds [the computer and automotive industries] collide.'

The big test, Balcombe knows, is likely to be public acceptance of the new technology. 'With Roborace, we've said there are three areas we need to focus on: develop the technology, develop the talent and develop trust. Trust comes from seeing what these technologies can do. And motorsport has always played a role in that. Think about Mercedes. The reason it is seen as a pinnacle brand is because of what they've been able to deliver in motorsport: it shows the quality of engineering they have is outstanding. I believe that the public being able to see AI systems engage in on-track combat, if you like, and do that safely and competitively is a key element to building trust in them. Yes, you could have an experience jumping in a robotaxi, but you're not expecting it to do an emergency manoeuvre. Wouldn't it be nice to know that, if needed, it can do that as capably as a car you see at a race weekend? That it can do incredible stuff you never realised was even possible?'

This comes back to di Grassi's annual race against the AI. Balcombe predicts that di Grassi isn't going to keep up his winning streak much longer. 'There's a one-hundred-and-seven per cent cut-off rule in Formula One,' he says, requiring competitors to be within a certain threshold of the fastest qualifier's time to race. 'The AI is within that [against di Grassi] now. It would pass the threshold for qualifying for a Grand Prix.'

There's a caveat, though: the races with di Grassi have been on dry tracks. For now, humans still have the edge when weather conditions are changing. 'That's the most challenging thing for a driver,' Balcombe says. 'It's where you see the likes of Lewis Hamilton excel. So, [an AI beating someone] on a single lap time in dry weather conditions? That's achievable within a year or two. But it'll be a long time before we see an AI system be able to take on Lewis across a Formula One season, when you have all the different variables throughout the year.'

And how does di Grassi feel about losing to a robot? 'My colleagues think I'm a turkey cheering for Thanksgiving,' he jokes. 'But AI is going to make us all unemployed anyway, we might as well accept it and develop it through motorsport. And remember, IBM's Deep Blue beat [then world champion] Garry Kasparov at chess in the 1990s, but that doesn't mean people don't play chess any more! People will always want to see the best driver, who's the most skilled and the most brave, who can drive best in the worst conditions.'

He grins as only a world champion can. 'I'm just glad I've had my prime. Because one day AI will be much better than humans. There is no limit to what AI can do.'

PART THREE

THE MATERIAL WORLD

CHAPTER TWELVE

Flax, Fibres and Floating Frogs

The periodic table is chemistry's road map. It was put together by Russian scientist (and bigamist) Dmitri Mendeleev in 1869, and is a handy guide for scientists to the building blocks of the universe. Most of the elements on the table are metals. These include ones you've probably heard of, such as iron, copper and gold, and ones you've probably not come across, such as erbium, terbium, yttrium and ytterbium, which were all named after a tiny village in Sweden called Ytterby.*

What counts as a metal depends on which scientist you manage to corner. A chemist, for example, would probably talk elements with crystal structures that form neat, ordered rows with tight covalent bonds. A physicist would tell you that a metal is any substance that conducts electricity at absolute zero (which, bizarrely, means some non-metals, such as iodine, can turn into a metal if you keep them under pressure.) And an astrophysicist would happily tell you that anything other than hydrogen and helium is a metal. But, then again, astrophysicists are weird.

Regardless of the definition you pick, the scientific consensus is that most car chassis are made from metal. As pure metals are usually very ductile and lack strength, this

* I've been to Ytterby – it's a short bus ride from Stockholm. Tiny is an overstatement: it's basically a few houses, a shop and a blocked-up old quarry. Even so, the quarry is one of the most important sites in modern science. Looks can be deceiving.

will almost always be an alloy, or a mix of different elements. Perhaps the best-known alloy is steel (a mix of iron and carbon, perhaps with a little chromium thrown in if you want to make it stainless), but there are a host of others – bronze, brass and pewter – depending on the properties you want.* Throughout this book, whenever I've referred to a metal used in construction, I've probably been referring to an alloy. For example, while it's true the halo device is made predominantly of titanium, it's actually titanium grade 5/6Al4V – a titanium alloy with about 6 per cent aluminium and 4 per cent vanadium. With literally millions of alloys to choose from, manufacturers will always use a material best suited to their needs. That means balancing strength, ductility, weight, conductivity, melting point, resistance to oxidation (rusting) and, of course, its price tag.

Until the 1950s, a Formula One chassis was usually made from a tube of aluminium surrounded by hand-beaten aluminium panels. In an industry that wants to stay as light as possible, though, alternatives are always considered, which is where composites come in: combining two distinct chemicals together to make a new material. In the early 1960s, Formula One began looking at glass-reinforced plastics and started using plastic inner layers. While this was a positive step, the plastic lacked the strength and rigidity to keep drivers safe. Ultimately, it was never going to work.

Then, in 1981, McLaren tried a new material for its chassis: carbon. This is, without question, the most versatile element

* Iron is one of the most common elements used in alloys. Its Latin name is *ferrum*, which is why its symbol is Fe. And the surname meaning 'blacksmith' in Italian? Ferrari.

in the universe. Open your eyes and look around: you're probably seeing carbon, whether it's the pages of this book, plants, rocks, plastics or … well, you. All life on Earth uses carbon as the backbone of its structures. It's in your DNA, sugars, fats and the walls of your cells. Everything we eat contains carbon and we exhale carbon dioxide. It's so important, one of chemistry's main branches, organic chemistry, is solely dedicated to 'carbon plus other stuff', with inorganic chemistry focusing on just about everything else. Want to confuse a chemist? Give them a plastic bottle and ask them whether it goes in the organic or inorganic recycling.

Carbon's versatility is explained by how it interacts with the world. It has four electrons in its outer shell, which means it can form up to four bonds with neighbouring atoms. And carbon *loves* to form bonds. This is why it's the perfect backbone for life on the planet: you can make molecules that are basically long strings of carbon atoms with a whole host of things coming off either side. And carbon–carbon bonds are covalent, meaning they give each other an electron, so are incredibly hard to break.

Thanks to its versatility, carbon can be arranged into a host of different shapes, called allotropes. This is how the same element can form something as tough as a diamond and as seemingly weak as the graphite in your pencil. But graphite isn't really weak at all. At a chemical level, graphite is just layers of carbon atoms, bonded together to form long sheets of hexagons. Each atom is securely bonded to three neighbours, with an electron left 'spare'.* While the links between these

* This is because of sp^2 hybridisation, but you really don't need to know that unless you're a chemist.

sheets aren't very strong – which is why a graphite pencil easily leaves a mark on the page – the sheets themselves are incredibly tough to break. Thus, when graphite is woven together in strands, the result is something that's low weight, stiff, highly resistant to chemicals and a great conductor of heat (about five times better than copper). It's a material known as carbon fibre.

Carbon fibres are nothing new. In the nineteenth century, they were used to make light bulb filaments. Remember that 'spare' electron? This is delocalised throughout the structure, something you normally see in metals, which makes graphite excellent at conducting electricity. And yet this was only the beginning. By the 1960s, fibres of graphite could be woven into thin strands, about 5–10μm wide, with a high tensile strength. By turning these fibres into a mesh and adding them to other materials, you could create a reinforced composite.

The result was a host of lightweight, strong materials that could do different things depending on what they were combined with. For example, McLaren's bodywork was carbon-fibre-reinforced polymer (CFRP), a plastic resin with carbon fibres woven into a mesh and trapped inside. This resin can then be cured and hardened, turning into a glassy, brittle solid that's fixed in whatever shape you need. But 'carbon fibre' can *also* mean carbon-fibre-reinforced carbon (CFRC), a completely different composite that uses carbon's thermal conductivity and temperature-resistant properties, making it perfect for the nose cones and wings of space shuttles re-entering the Earth's atmosphere. In fact, by the time McLaren started experimenting with their bodywork, CFRC was already being used in Formula One: it had been part of the cars' brake discs since 1976.

If the idea of tough, heat-resistant fibres is reminding you of the aramids we discussed in Chapter Ten, you're right: they work in a similar way, just without the amide bits linking them together. And today, much like aramids, carbon-fibre composites are everywhere in motorsport, offering strong, lightweight alternatives to metals across a car's design. It's no surprise that Formula One teams have even used their expertise in the material to branch out beyond racing. Carbon fibre is already widely used in aerospace but, in 2019, Williams Advanced Engineering partnered with JPA Design to take things even further. Working with two CRFP composites it developed – Racetrak and 223 – the team looked at whether it could use them in plane interiors. The result was a new type of plane seat that got rid of the aluminium frame altogether, shaving off 4kg and making it about 5 per cent lighter than the existing versions you'll find in passenger airliners around the world. While 5 per cent doesn't sound like a lot, the potential impact is huge. Airline traffic is responsible for about 2 per cent of emissions globally. According to the FIA, if you took just 12 long-haul planes and replaced only the seats in business class with the Williams design, it would cut 942,000kg of CO_2 emissions a year and save about £140,000.

Unfortunately, carbon fibre has some serious flaws. It's expensive to produce and making it isn't exactly environmentally friendly. Today, carbon fibre usually comes from oil, where it's turned into fibres of a plastic called acrylonitrile (CH_2CHCN). This then undergoes calcination, which involves heating at temperatures of more than 1,000°C in an oxygen-free environment; without oxygen, the fibres can't burn and eventually you're left with strands of the carbon. The process means around 20kg of CO_2 is produced just to make 1kg of carbon fibre. And while the weight savings mean that, when used properly, carbon fibre cuts down CO_2 emissions and saves money in the long run, it'd be much better for the environment if there was an equivalent material that didn't have such an impact in the first place.

It's a good thing carbon's so versatile, then. Such a material already exists – and it's made from the same thing as your bed sheets.

* * *

At the tail end of 2020, Kyoto University professor and former astronaut Takao Doi proposed an ambitious new material for satellites: wood. The satellite could go up, do whatever it was built for, and then be programmed to fall back down to Earth, where its wooden frame would burn up harmlessly on re-entry. This, Doi argued, would solve two problems. First, it would prevent the Earth's atmosphere being surrounded by space junk. (Already, there are an estimated 129 million pieces of debris in orbit, ranging from microscopic paint flecks to about 34,000 chunks of metal larger than 10cm, all whizzing around at 22,000mph, ready to smack into anything that gets in their path. It'd be much better if we could clean up after ourselves.) The second problem Doi noted was that, while we know what happens when we burn wood, we have absolutely no idea what will happen to our atmosphere when metal satellites burn up on re-entry, scattering their alumina particles across the sky. It might be harmless, or we might be damaging our planet in some way. Why take the risk?

Although this sounds smart, there are problems with making a wooden satellite. For starters, it would give off volatile organic compounds (aka tree smell), which could impact your instruments. It wouldn't conduct electricity, which means it can't help ground any electronics from built-up charge. Nor would it dissipate heat, protecting the craft from the vast temperature changes in space. Finally, wood also contains moisture, which would evaporate in space and compromise the satellite's structural integrity. But, if you think these flaws mean plants have no place in high-end space tech, you can think again. Plant-based satellite panels already exist – and come from a technology that emerged through racing.

In September 2020, Porsche launched a Cayman 718 GT4 car at the 24 Hours Nürburgring in Germany. It looked like any other Cayman but under its paintwork it had a secret: its bodywork was entirely made from interwoven flax fibres. 'It's the pinnacle of our development with Porsche,' says Johann Wacht, manager of motorsports for Bcomp, the company responsible for the feat. Growing up near the Nürburgring itself, Wacht has always been fascinated with cars. Now he's seeing his work on the track. 'Our material has the equivalent stiffness and weight to carbon fibre, within a certain performance window. But we save approximately 85 per cent of its carbon footprint and can cut raw material costs by up to 30 per cent.'

Making things out of flax is far from a revolutionary idea; woven flax is linen and has been used by humans for about 30,000 years. It's so old, it's even referenced in the Bible.

Humans have used flax for rope, for canvas, for teabags. We even use 'flaxen' to describe blonde hair. So why did Bcomp win the 2018 World Motorsport Symposium's award for most innovative product of the year? The secret, like carbon fibre, isn't the material. It's the design.

'The basic concept is inspired by nature,' Wacht says. 'We use a thin surface on one side and, on the other, a 3D grid called PowerRibs, inspired by leaf veins. Without the ribs, you just can't compete in terms of weight or stiffness with carbon fibre. The ribs are the underlying magic that brings the performance.'

Bcomp's flax-based tech has major advantages. The first is sustainability – not just in terms of reducing carbon footprint, but by working within current farming practices. 'Flax is a very old, established European crop,' Wacht says. 'It's actually used as a fallow crop to re-energise soil. In Belgium and France, farmers plant flax every three to seven years. So, we have a mature supply chain in place.' Not only are natural fibre parts as easy and cheap to make as carbon fibre, the fibres are also decomposable (although the epoxy resin used to bind them into place means the material isn't completely recyclable). Rather than go to landfill, they could also be burned as an energy source.

Flax also has respectable safety credentials. While some carbon-fibre composites are designed to be used in crash structures, others on non-essential parts of the car aren't as tough. During a collision, the epoxy used for these weaker carbon-fibre wings often breaks off, shattering the bodywork and leaving sharp, spiked shards on the road. 'In a crash, you have a big risk of splinters all over the track,' Wacht says. 'That creates the risk of a puncture, causing an even bigger crash.

So, in 2019, GT4 introduced a rule that required aerodynamic parts to have a mandatory amount of natural-fibre composites [in place of carbon fibre]. Porsche had already gone further – all of the doors of the current Porsche GT4 racing cars are natural-fibre composites. After that, Audi, KTM and Toyota also joined the club.'

Aerodynamic wings are only the start of flax fibre's potential. 'The next logical step is to look at where we can push the technology further, and that's structural applications,' Wacht says. Already, Bcomp has passed crash tests using its flax, showing that it's a safe alternative to carbon fibre for a car's safety cell. The downside is that flax fibres are still too heavy. 'The prototype was about 40 per cent heavier than carbon fibre at the moment,' Wacht says. 'That sounds like a lot, but when you're looking at the overall weight of a car, it's less than 1 per cent. We need to ask whether we value speed or sustainability. Do we really need to have a carbon fibre crush box, or could we have just a tiny bit more weight and take out 40–50 per cent of the material's carbon footprint?'

Given its green credentials, Bcomp is already attracting attention from Formula One. 'McLaren is one of the first teams that looked at sustainability before it was cool,' recalls Wacht. 'They'd been looking for a holistic approach for years to go carbon neutral, and decided one of the parts of the car where they could introduce natural fibre composites was in seats.'

As mentioned earlier, Formula One seats are designed to be removable. To make this fair for all teams, the rules require that the seat and driver combined must weigh at least 80kg. Most drivers, including McLaren's Lando Norris, are far lighter, so often their seats need ballast. 'Usually, they use carbon fibre, which has the highest performance, but it's just waste material,' Wacht explains. 'It doesn't make sense from a sustainability point of view. So, we've worked with McLaren to use natural fibre technologies in the drivers' seats, taking a massive amount of CO_2 waste out of manufacturing the part. Natural fibres also have another advantage: they dampen vibrations in the seat [making things more comfortable for the drivers].' Natural fibres only became legal in Formula One in 2021,* but McLaren is already looking at how much more of its car could be made from, well, linen. 'McLaren is saying there's loads of potential for natural fibres in F1. But, to be honest, the bigger potential for natural fibre isn't in cars. It's in the rest of the team's eco footprint: timing stands, pit gear, all of the stuff that goes on away from the track.'

Bcomp has already taken the skills it's learned on track and applied them to space, building fibre-based satellite panels for the European Space Agency. And larger projects are possible, too. In 2020, the team worked on creating a 66m long, 4m-wide bicycle bridge near Leeuwarden in the Netherlands. Around 80 per cent of it was built from bio-based materials, mixing Bcomp's flax with balsa wood. Expected to stand for 50 years, the bridge weighs only a third of a concrete equivalent and is CO_2 neutral over its lifetime.

But while natural fibres are a terrific option for some projects, there's yet another material that's soon going to rock

* The regulations now allow for 'flax, hemp, linen, cotton [and] bamboo' to be included on an F1 car.

our world. A few years ago, I asked one of the world's most famous organic chemists, Professor David Leigh, what he wished he'd known at the start of his career. 'Well,' he sighed, 'I wish I'd known I could win the Nobel Prize with some Scotch tape and a flippin' pencil.'

He was talking about the discovery of graphene.

★ ★ ★

Andre Geim levitates frogs. In 1997, bored one Friday night, the physicist decided to see what would happen if he poured water on his laboratory's electromagnet while it was going at full power. 'Apparently, nobody had tried such a silly thing before,' Geim later recalled. 'To my surprise, the water didn't spill on to the floor but got stuck in the vertical bore of the magnet.'

Messing around with a colleague, Geim altered the settings and broke up the water with a wooden stick, managing to create small water droplets that seemed to levitate. Soon, the duo realised their floating water was down to an effect called diamagnetism – a very weak force that counterbalances a magnetic field. Usually, diamagnetism is so small it's insignificant. But, if you have a really powerful magnet and a really light object, you can eliminate gravity.

Geim decided to see if he could make a living thing levitate. He picked out a frog from the biology department, started up his electromagnet and soon had his amphibian test subject levitating helplessly around the bore. In 2000, his research paper on how to do it – 'Of flying frogs and levitrons' – won him the not-remotely-prestigious Ig Nobel Prize, a joke award given to scientists who've come up with the silliest piece of work that year.*

Geim's motto is, 'It's better to be wrong than boring', and, even before his floating froggos, he'd lived a life that probably qualifies him as one of the most interesting men in the world. Born in Sochi, Russia, now the host of its own Grand Prix, Geim had an eccentric entry into science. His family was of German descent and most of his relatives, including his father and grandfather, spent time in Soviet gulags. After graduating in the top 5 per cent of his school, Geim applied to the Moscow Engineering and Physics Institute and was shocked when he failed his exams. A year later, he tried again – and failed again. It was only then he realised he was in a room with 20 other candidates, all with Jewish or foreign-sounding names, and they'd all failed as well. Racism in Soviet Russia was alive and well. Undeterred, he instead applied for the Moscow Institute of Physics and Technology – the best university in the country – where he graduated in the top 5 per cent of his class. From there, he became a lieutenant in the Russian Red Army, studied intercontinental ballistic missiles, climbed five mountains, worked as a bricklayer in the Arctic Circle and survived a 100m fall down a crevasse without a rope. Somehow, he also found time to complete a PhD. In 1990, he moved to the UK, then on to Nijmegen and his flying frog.

* Other Ig Nobel prize-winners that year included a device that detects if a cat has walked across your keyboard and a peace prize for the Royal Navy, which had ordered its sailors to stop using live shells during practice and just shout 'Bang!' instead.

But while floating amphibians are important, it's only the second-greatest accomplishment to emerge from his fevered brain. After the frogscapade, Geim had taken his 'Friday night experiments' and turned them into regular a group activity. In 2004, bored one Friday night in Manchester, UK, Geim gave a lump of graphite to one of his students, Da Jiang, and asked him to polish it so he could study its properties.

Thanks to the language barrier – neither of them being native English speakers – Jiang misunderstood. Instead of polishing the graphite, he ground the sample down until it was little more than a tiny flake at the bottom of a petri dish. Geim found the result hilarious. He teased Jiang, joking that he'd 'polished a mountain to get one grain of sand,' and began to instruct him on what he actually wanted. As he did so, a visitor to the lab from Ukraine, Oleg Shklyarevskii, happened to be standing nearby. He overheard and, deciding to help, fished a strip of discarded Scotch tape out of a nearby litter bin.

Shklyarevskii was an expert in scanning tunnelling microscopy and graphite was used as a reference sample to calibrate his machines. The normal way to get a nice, fresh surface of graphite, Shklyarevskii explained, was to rub sticky tape on it. The tape could then be whipped off, leaving a nice, smooth patty. If Geim was looking for a thin layer of graphite, he could help himself – it was embedded on the Scotch tape they'd been throwing away for years.

Geim did something no one else had ever considered. He took the piece of tape, a tiny layer of graphite still smeared on its surface, and stuck it under a microscope. 'Polishing was dead,' he later joked. 'Long live Scotch tape!'

His investigation soon paid off. As we know, graphite exists in loosely bonded sheets. What Geim managed to do was obtain a *single sheet* – a one atom thin, almost two-dimensional layer of carbon, bonded tightly together. As he studied it, he realised he'd made perhaps the greatest discovery of the twenty-first century. A 2D layer of graphite had incredible properties. It was lightweight, able to conduct electricity and temperature, and almost impossibly strong. Better yet, it was cheap, plentiful and easy to make. It was the discovery of graphene.

Geim submitted his work to the journal *Nature*, whose reviewers rejected it as it did 'not constitute a scientific advance'. He submitted it again and was rejected again. Finally, he submitted it to the journal *Science*, where the reviewers realised just what Geim had discovered. Six years later, thanks to a simple misunderstanding and a bit of tape fished out of the trash, Geim got another award to go along with his Ig Nobel trophy: the 2010 Nobel Prize in Physics.

★ ★ ★

Graphene is being hailed as the wonder-material of our age. It's something that's so abundant we've all been using it for years without realising, and yet it's causing the biggest materials transformation in 100 years. 'We learned how to arrange carbon atoms in a string about a century ago,' explains Professor Vincenzo Palermo, a director at the National Research Council of Italy. 'That ability, to arrange molecules in one dimension, created the plastics revolution. What we're doing with graphene is producing materials in two dimensions – a honeycomb lattice. It's a sheet, not a string. And with a sheet, you can do a lot of different things.'

Graphene's honeycomb shape and carbon–carbon bonds mean that, from a mechanical point of view, it can withstand tensile stress from any direction as its hexagon can deform. This makes it very tough and robust. But, being a sheet, it's also easy to bend into any structure you want. 'Farmers knew this before scientists,' Palermo says. 'Think about chicken wire: it's the same principle. You can bend it easily with your fingers but if you want to break it, that's tough.' The hard part to imagine is that this sheet is only *one atom* thin – about a billion times smaller than the palm of your hand. And yet, Palermo continues, the sheets are just that: a sheet. They might be thin but they can stretch out millions of times wider than their thickness. 'You can have sheets about the size of a human hair – so you can see it in a microscope and barely see it with the naked eye. It blows my mind! Physicists tell us you can't see something smaller than the wavelength of light. But, in this case, you can! And it's mechanically stable!'

This isn't the only weird thing about graphene. Its delocalised electrons behave as massless particles. 'This makes it the perfect playground for physicists,' Palermo says, 'because they can see all kinds of strange quantum physics phenomena, even at room temperatures, using something that you can get with a piece of Scotch tape.'

Unsurprisingly, motorsport teams are already looking at graphene in all aspects of the car. You could build the body of the car out of composites based on graphene, making it thinner and lighter while retaining its strength and safety. It's so good at conducting electricity that you could shrink your circuit boards and wiring, reducing the (literal) miles of cables that go into the cars. 'We think the properties of graphene are pretty mind-blowing,' McLaren's chief operating officer, Richard Neale, said in an interview on the team's website. 'Some of the mechanical properties of graphene-enhanced composites can be improved by double-digit percentages compared with regular carbon-fibre composites. In engineering, we talk about improvements in terms of fractions of a per cent; to suddenly introduce improvements of this order is incredible.'

Graphene has already started to appear in sport. It's been used in tennis racquets and in motorcycle helmets with graphene flecks as an added coating. But, in truth, that's not where graphene is going at all – you're thinking too big. Remember its incredible ability to conduct temperature? If you put just a little bit of graphene into a fluid, you can make it an incredibly effective coolant for engines, batteries, computers and just about anything else that needs heat drawn away from it. Already, graphene-based coolants are about 60 per cent more effective than the alternatives. As we explored earlier, temperature is often the limiting factor for motorsport. If you can make a more effective coolant, not only can you use ever-more powerful systems, you can also rethink their shape and layout. Better still, graphene coolants could solve the tricky problem of temperature spikes when recharging electric vehicles – meaning that ever-faster, more convenient charging is on the horizon. And when you combine that with graphene batteries and graphene electrical circuits, you're suddenly able to store massive amounts of energy and transfer it in an instant, all while keeping the temperature cool. You could charge a car in minutes, or your mobile phone in seconds. These chargers won't have to be large and bulky either; in 2008, Geim and his team managed to create a graphene transistor one atom thick and ten atoms wide. Graphene is also being used to develop flexible, stick-on sheets of solar cells capable of turning any surface into a power station. Your super-speedy, super-capacity batteries will be charged by the sun.

Graphene's applications don't stop there, Palermo says. Its origami-like ability to fold into whatever we need means it is likely to affect every aspect of our lives. We could use it to create smarter, better-targeted medicines, ultra-sensitive biosensors to monitor diseases, or as nanoscale bandages to repair wounds at the cellular level. Graphene oxide can be used to create food packaging that's able to detect whether something is going off, all while keeping out oxygen and moisture – a tiny, single-atom-wide net that won't let anything through. And thanks to carbon's versatile nature, the material can be both water-loving and water-hating, making it the perfect filtration system to purify drinking water. With the UN predicting 14 per cent of the world could be living in water scarcity by 2025 due to climate change, the importance of being able to use *scrapings from a pencil* to provide safe drinking water for millions can't be underplayed.

Put simply, in the same way a Victorian wouldn't be able to imagine how we use plastic, we can't even begin to picture what our world will look like after the graphene revolution. Just don't expect it to happen all at once. 'We often tell a false account of how things work,' Palermo says. 'That there's this Eureka moment and then everything is very fast and very easy. It's not. Most new technologies take twenty to forty years. Think about plastics, or the microchip. Think about how the electric light bulb took decades to get to where we know it. Graphene is going to take time. You'll see commercial products appear and you'll see graphene in sensors. But we're not expecting to build a space elevator or things like that any time soon.'

We're still going to see it in motorsport in the next two to three years, though. 'The first people that get on to any new technology are always the military and sports teams,' Palermo says. 'Your typical person might not be willing to pay an extra £500 to shave off ten grams from their bike, but armies and elite sportsmen and sportswomen are. You're already seeing companies like Vittoria use graphene in their bicycle wheels and tyres.' The graphene revolution might be far closer than we think.

I'm reminded of the words I heard back in Valencia. *Sport and war. That's where most innovation comes from, right?* And there's another area of materials science this dual-headed dragon of human competition has its eyes on, too. In a sleepy, rural stretch of Tennessee, the US government is already creating racing cars, attack helicopters and submarines out of thin air.

CHAPTER THIRTEEN

All That's Fit to Print

The Shelby Cobra is a work of effortless cool. I'm sitting in the driver's seat of a legend – finally, a car that fits! – parked on the second floor of a building in Tennessee. If I started her up, I could make 0–60mph in about 4 seconds. Although, judging by where I am, I'd probably smash through several million dollars' worth of US government property first. The Cobra is inside Oak Ridge National Laboratory's Manufacturing Demonstration Facility, or MDF. A fairly innocuous complex away from the lab's sprawling, secretive campus just outside Knoxville,* it was built in 2012 to be the US Department of Energy's premier playground for materials.

There are few cars more instantly recognisable than the Cobra, especially in Shelby American's classic livery of guardsman blue with two Wimbledon white stripes sweeping over the centre of the bonnet. Between 1965 and 1991, it was the most powerful production car on Earth. It was created by the great Carroll Shelby, the chicken farmer who won the 24 Hours of Le Mans and then partnered with Ford to break Ferrari's dominance of the race in 1964. This particular Cobra is special, although when it was showcased at the Detroit Motor Show in 2015 most visitors just walked past, thinking it was a normal classic coupe.

---

* See my previous book, *Superheavy*, for more about Oak Ridge's top-secret history and atomic links.

Nobody realised the entire car had been 3D-printed from scratch.

'We had actually printed cars before,' says Alexander Plotkowski, a staff scientist and my guide for the day. 'The problem was that they *looked like* 3D-printed cars – they have that kind of rough, layered look to them. We did the whole Cobra in about six weeks and finished it so that it's indistinguishable from a conventional car.' Everything on the Cobra, save the tyres and the engine block itself, came from the lab's machines, squirted like superhot toothpaste from a computer-controlled extruder. Inside, you can see the telltale ridges and bumps where the surface hasn't been completely polished down. If you passed it on the road, however, you'd never know. 'We haven't printed a complete engine, we're not there yet,' Plotkowski says, 'but we're working on it. It's only a matter of time.'

The MDF isn't your typical lab: it's the 3D printing equivalent of Willy Wonka's Chocolate Factory, filled with so many astonishing things it's hard to process them. The Selby Cobra is just there to whet your appetite. It's also the last point in the building I'm allowed to use my camera – everything else is top secret.

Out of the Cobra, the tour continues into the main hub of the MDF, a 10,000m^2 open-plan warehouse filled with industrial machines – forges, electron microscopes and CT scanners – and finished by a massive 20ft US flag on the back wall. While it's owned by the US government, most of the equipment is donated from companies such as Boeing, Lockheed, Honda, General Motors and Chrysler. Taking up one wall is a white sheet, behind which the lab is 3D-printing prototypes for a nuclear reactor core. With everyone using the same tried-and-tested reactor patterns since the 1950s, they figured it was time for an upgrade. On the other side of the workspace, looking like a greenhouse on stilts, is the lab's Big Area Additive Manufacturing (BAAM) unit. In 2017, the lab used it to 3D-print a minisub for the US Navy. This wasn't the lab's longest or most complicated project: a single part for a helicopter once took three weeks to print. Nor is it their biggest printer; if the team runs out of space, they can access a system at the University of Maine capable of spooling 200kg of material an hour to create objects 30m long, 6m wide and 3m high. That's equivalent to printing two train carriages side by side.

'What's that?' I ask, pointing to a big, curved *something* on a plinth. It looks like a rotor, or a propeller, or maybe a weird art installation.

Plotkowski hesitates. Glances around nervously. 'Ah … I can't tell you what that is,' he says slowly. 'It's OK for you to look at it, as long as you don't know what it is. If you did know what that was, you wouldn't even be allowed to look at it. All I can say is that's it's to do with … uh … aerospace.'

It is literally an *unidentified flying object.* I nod conspiratorially and the tour moves on. But even if I'm still in the dark about whatever secret aerospace project the US government is up to, at least I know I'm in the right place if I want to understand why 3D printing is the next giant leap in design.

* * *

Additive manufacturing – the proper name for 3D printing – has been around since the early 1980s, but it's only really taken off in the past 10 years. Visit any top university today, and you're likely to find 3D-printer stations for students to use, turning whatever design they imagine on screen into reality. Although there are several different techniques and methods, essentially, you upload your design into a computer and programme it with the precise geometry your shape needs. You add your feedstock into the 3D printer and the computer does the rest, squirting out your shape, one layer a time, in a real-life version of a replicator from *Star Trek*. Thanks to the computer's accuracy and precision, this means you can build highly detailed, unusually shaped parts quickly and easily. And, as you are building the object from the inside out, you can accomplish feats of engineering that wouldn't be possible with conventional moulds. Better still, because you're printing only what you want, you cut down on waste materials.

Typically, additive manufacturing involves either plastics or metals. For the former, this means you usually need a binder that can glue the layers together. Currently, the most common is furan, C_4H_4O, although Oak Ridge is already looking at alternatives given this is horrendous for the environment. This is a type of 3D printing you can do at home and is basically a computer-controlled glue gun. While just 10 years ago these had relatively low resolution, now you can print incredibly complicated shapes and designs. Today, you can even buy Halloween masks that have been created in 3D-printed moulds and custom-made to your exact designs and specifications.*

* I was going to add in a few paragraphs about my own experience with this, when I used 3D printing to create a mask that turned me into an eerily lifelike double of Emilia Clarke from *Game of Thrones*. Apparently, though, people think it's a bit weird to run around wearing the silicone head of a famous actress, even for Comic Con, so that part's been cut from the book. Today, the fake Emilia head is kept at my mum's house with a pillowcase over it, so as not to freak out any visitors.

For metals, which have higher melting points than plastics, this process needs to be done at much higher temperatures. Fortunately, no binder is required: you're just telling a computer to do thousands of micro-welds a second, a process known as 'sintering'. Inside the furnace-like 3D printer, lasers and electron beams liquify the structure as the next layer is sprayed on, with the metal deposited as microscopic particles trapped within streams of inert gas. This requires finesse; printing creates temperature gradients, which can cause parts to crack. For quality control purposes, you also have to be certain that, if you use a machine 100 times, you're going to get 100 more-or-less identical parts. Getting 3D printing right is a billion-dollar research project ... although it could lead to trillion-dollar solutions.

Unsurprisingly, motorsport was an early adopter of 3D printing. Every Formula One team uses additive manufacture to create new parts, allowing them to prototype designs rapidly and cheaply to see what works; the FIA estimates it reduces production times by 20–25 per cent. As The Ohio State University's David Cooke explains, this works in tandem with simulations. 'You run a thousand cases in a computer, you narrow it down to the three you want to try. You 3D-print some parts, and suddenly you know in two days what would have taken two years not too long ago.'

Prototyping is just the beginning, however. Ferrari has used the technology to create parts for its power unit, McLaren 3D-printed the hydraulic line bracket of their 2017 car and the Renault F1 team's cars for 2019 each contained around 100 3D-printed parts. Alfa Romeo Racing's parent company, Sauber, even has its own propriety brand of carbon-reinforced powder specifically for printing, which it uses to create small parts for its aerodynamic set-up. Another major investor is BMW and its Formula E team. In 2020, not only did the BMW iFE.20 racer use flax fibres for its cooling shafts but it used its own facility near Munich, Germany, to 3D-print a 360-degree aluminium motor casing. In 2019, BMW 3D-printed around 300,000 parts, showing just how keen it is to adopt the technique.

It's not just car manufacturers that use the process, either. In 2006, Michelin began looking at 3D-printed machine tools. Michelin knew the shape of the tyre it wanted to optimise grip, but didn't have a mould that could make it. Instead, it reverse-engineered the design, figuring out what the mould needed to look like. Then, it partnered with a precision manufacturer, Fives, and invested millions in developing machines that could 3D-print moulds to create the treads it wanted. In 2017, the company extended 3D-printing to its tyres. This led to its creation of an 'airless' tyre, replacing the inflated inner tube with a honeycomb structure created from a mix of cardboard, plastic, tin cans, tyre chips and orange zest.

It looks like 3D printing is here to stay. And yet this barely scratches the surface of what 3D printing can do – or how it's already being used to revolutionise the motor industry.

★ ★ ★

Past the mysterious top-secret-UFO-thing, Plotkowski leads me into an area filled with 3D printers, each one currently in operation, monitored by thermal cameras to ensure temperature is even and the parts aren't liable to crack. Walking over to a table with samples littered on it, he passes me something instantly recognisable to any petrolhead: a turbocharger.

'When the 3D-printing tech was in its infancy,' Plotkowski says, 'everybody would just say, "I have this widget and we're making it by forging and machining at the moment. I want you to print it with the same material, the same property, the same shape." And that's not a very useful thing to do – because you're saying you want the same thing, just 100 times more expensive.' Instead, he gestures to the turbocharger. 'But this is a case of us designing an aluminium alloy from the ground up. It actually has better properties than cast aluminium. Not to get super-technical but, in conventional casting, the cooling rates are relatively slow and that means it has poor microstructures. So [historically] we've had to design heat treatments and processes to get the strength up. Now, in additive manufacturing, the cooling rates are extremely high. So, I can design an alloy that takes advantage of that to gain strength. The microstructure of this alloy looks dramatically different [from conventional aluminium] – and we've designed the chemistry to take advantage of that.'

This gives engineers even more options to play with when choosing the right material. You can make parts for cars that are lighter and thinner, yet also tougher and more reliable. You could also purpose-build things that have better electrical conductivity, or are more flexible. Plotkowski is leading the aluminium project (although he insists on calling it 'aluminum', which wounds my British soul). 'We've been talking with racing teams to do some early-stage testing of the kind of components that can be printed,' he says. 'Our goal is to print high-temperature aluminium pistons that have better performance than existing ones. Now, the Department of Energy isn't interested in racing cars. But the racing teams are where big manufacturers can spend money on risky investments, work things out and then we can adopt these technologies on scale much more easily. 3D printing is a new technology that the big automotive manufacturers aren't ready to adopt yet, but they're all talking about it – and how do you demonstrate your part has some benefit?'

'By making it go around a track at two-hundred miles per hour repeatedly,' I finish. Plotkowski nods. From lab, to racing car, to road.

We're just getting started. Walking further into the metallurgy section, I'm brought towards a large machine from Haas. 'So, this is a regular mill,' Plotkowski says, pointing at how it's grinding an engine block away. 'But I can also switch it over to deposition [of material], so I can add and subtract – build parts on the fly and then smooth and finish them. I can also do some really unique things. Say you have an engine that's been in operation for several years with a crack in it. I can identify the location that has a problem, deposit material [to fill the crack], then finish it and put the engine back into service.' While that might not sound exciting, what Plotkowski's just said is revolutionary. Currently, there's a billion-dollar market in rebuilding engines because no two cracks or faults are alike and each has to be repaired in a different way. Usually, once an engine breaks down it's cheaper just to throw it away – a huge waste. With one machine, you can identify the problem, fill the hole and then smooth it down from *inside the engine* to make it as good as new. An astonishing amount of waste is eliminated and millions of dollars saved.

The variety of shapes you can create with 3D printing – beyond the ability of conventional tooling – is yet another advantage. Usually, if you want to build something complex you have to weld two different parts together, which creates a point of weakness. With 3D printing, you can just tell the computer where you want it to leave gaps. Plotkowski leads me over to the side of the table, where a large, articulated claw is resting. 'This is a project we did with the Office of Naval Research,' he explains. 'It's an underwater robotic arm. There's a couple of really cool things you can only do with 3D printing – one of them is that these channels inside are actually hydraulics, there's no tubing.' Hydraulics are notorious for going wrong; in one print, the Oak Ridge team has eliminated that risk by making the hydraulic system as an integral part of the arm's structure. 'The inside of the component is also hollow,' he adds, 'so that even though it's made out of titanium, it's neutrally buoyant.' He takes a piece of metal and passes it over; it feels as light as holding a cup of coffee. 'Being able to operate this underwater suddenly becomes quite a bit easier.'

While lightweight submarine arms are cool, they're still relatively modest compared with some of the MDF's ambitions. Plotkowski leads me into a far corner of the lab, a large, seemingly empty space. Above it, a 3D printer is zipping around on wires, like a camera at a football game. This is where the lab is looking to press beyond the limits of even their giant machine up in Maine: it's a 3D printer with no size restrictions. Once operator Brian Post has set up his wires, he can make the 3D printer move anywhere within it – like a spider hanging from its web. I am *inside a 3D-printing robospider's web.* Science is cool.

'At the moment, if you're going to make a building, you need a printer that's bigger than the building,' Post explains. 'The problem is that's cost-prohibitive, requiring a giant gantry system and level terrain. [Instead], I can set this up, do all the inverse kinematics and know exactly how to move this around to make different objects, and then I can print what I want.'

'Like an entire house?' I ask.

Post nods. 'Most construction right now is linear; we make things that are in straight lines because that's the shape building materials come in. Here, we can make complex curvature. We can print architectural features, like complex columns or spiral staircases, at a cost nothing more than it would be to make a straight wall.'

'So, if I wanted to live in a giant statue of a cow, you could print that?' I ask, wishing I could somehow claw the words back into my skull. I have no idea why I thought of that.*

'Certainly,' Post says, with a nervous laugh and a quick glance at Plotkowski to ask why his guest came up with something so bizarre. 'You could live in whatever you want, subject to the design limitations of the material.'

While bovine-themed housing might be a tad niche (and, let's face it, weird), the tech has more practical uses. Already, the US Marine Corps has used the concept to build a base, 3D-printing an entire 46m^2 concrete barracks in less than two days. Typically, the Marines would put up a wooden structure in five. 'Robots should be doing everything that is dull, dangerous and dirty,' said Captain Matthew Friedell, in charge of the project. 'And a construction site on a battlefield is all of those things.'

---

* I later realised that I was thinking of the end of the TV show *Pushing Daisies*, where one of the characters opens a restaurant shaped like a giant cow. Unfortunately, that realisation dawned a little too late for me to save face.

The MDF has already come up with another, more peaceful use for the technology: disaster relief. 'We're looking at bio-derived materials as well, such as natural cellulosic fibres,' Plotkowski says. 'Imagine there's a natural disaster in Puerto Rico and you want to print telephone poles. Well, something that Puerto Rico has in abundance is bamboo, it's the perfect climate for growing that. So, we could send down a 3D printer and we could make those fibres on site [out of bamboo], mix it with whatever low-cost polymer they like in the right proportions and print whatever they need.'

It's an incredible thought. If a disaster strikes, rather than sending tonnes of aid, all that's required is a 3D printer, a few wires and poles and enough resin to glue your bamboo together. Rather than wait months for infrastructure to recover, a small team could get power lines and communication back up in a matter of hours, as well as printing resilient, basic shelters for survivors. Sure, it might be rough-and-ready, but it could save lives and provide a temporary solution until full repairs are possible.

This realm of untapped potential is the perfect place to end my tour of the MDF and the next generation of manufacturing. But the US Department of Energy isn't just working on new materials that could find their way into motorsport; it's also coming up with alternative ways to power the racing cars of tomorrow. It's time to consider our options beyond fossil fuels.

CHAPTER FOURTEEN

Fuelling the Future

You've probably not heard of Emeryville, California. It's sandwiched between the University of California, Berkeley and the metropolitan hub of Oakland, and doesn't even have its own station on the Bay Area's transit network; instead, you've got to make your way from the twisting routes of the MacArthur Maze, a sprawling, spilling spaghetti pile of a junction that funnels traffic across to San Francisco or down to the docks. It's sleepy enough that it runs its own shuttle service, the Emery Go-Round, to bring any visitors to the companies that call it home.

Riding the free bus, it's not long before I'm passing the bright gates and manicured lawns of Pixar Studios. Sadly, animated movies aren't in my future: my destination is just around the corner at the Joint BioEnergy Institute (JBEI), part of Lawrence Berkeley National Laboratory. I'm here to talk about alternative fuels.

While electric cars represent the best option for your daily commute, there are still inherent problems. First, you've got to charge them somehow. Second, they only operate in a narrow range of temperatures. Third, the components used to make batteries, such as lithium and cobalt, are finite resources; eventually, there's not going to be enough to go around. Most crucially, they still have a limited range per charge, which means they're not suitable for continuous long-haul trips. You can't just make a bigger battery either, as this will add mass. The classic example here is trying to create a battery-powered airliner. If you wanted to get across the Atlantic Ocean, you'd need a battery so large the plane wouldn't be able to get off the ground.

It would be great if you could just recharge your battery on the move. While this can be done on a small scale with regenerative braking (see Chapter Four), it's not possible to do it completely. Solar power is also still in its infancy. While solar cars are available, the recharge time – about an hour to get enough charge to go 7.5 miles – means they're limited to small-scale commuting or, more likely, as a secondary power source to give you a little extra juice. Racing has been at the forefront of trying to develop solar cars for years, most notably in the World Solar Challenge across Australia.* But, until the sun's rays can handle anything more than small, light vehicles, they fall outside the scope of this book.

There's also the problem of what to do with the estimated one billion vehicles in the world still operating on conventional ICEs. Ideally, in the short term at least, you want a green alternative that can be used in these existing engines. This is why a lot of research has gone into biofuels – energy produced from plants.

Biofuels have been with us from the start, when Palaeolithic hunters threw logs on to the fire to keep themselves warm. The first oils were also biofuels, either taken from plants (such as olive oil) or animals (such as whale oil). By the early twentieth century, the first generation of biofuels as an alternative to

* There's a cheesy 1990s movie about this starring James Belushi called *Race the Sun*. Among the before-they-were-famous supporting cast are two Oscar winners, Halle Berry and Casey Affleck, as well as Joel Edgerton and Eliza Dushku. It's basically *Cool Runnings* only with solar cars.

petroleum had emerged. These were typically from food crops, taking the sugar, starch and oil from plants such as sugar cane and corn, and fermenting it to produce ethanol (C_2H_6O, the same alcohol as you'll find in beers, wines and spirits). Sadly, booze-based biofuels weren't financially sustainable as a crop. Next came the so-called second generation of biofuels. These use waste or by-products from farming, taking the residue to make fuels. These are roughly where we're at today.

Biofuels still release carbon dioxide when they're burned, but at a much smaller net release than fossil fuels; remember, as plants grow, they take in carbon dioxide from the atmosphere as part of photosynthesis, helping to offset their overall carbon footprint. Unlike fossil fuels, they are also a renewable resource: one day fossil reserves will run out, but you can always plant more crops. The downside is that you get less bang for your buck: for every 1 gallon of petrol, you need 1.5 gallons of ethanol to keep your engine running. Despite this, biofuels currently support 3 per cent of the world's road traffic, either as the sole source of energy or blended with regular petrol. Some countries have gone even further; around 25 per cent of vehicles in Brazil run on biofuels in some form.

But this is still barely scratching the surface of what's possible. 'There isn't a silver bullet for climate change, no one solution that can fix it all,' says Aindrila Mukhopadhyay, a senior scientist at the JBEI. 'But having biofuels at scale could be part of that solution.' Originally a chemist who studied how microbes sense their environments, Mukhopadhyay turned her skills to taking renewable feedstocks and making the next generation of biofuels. She uses algae.

These third-generation biofuels are only just emerging and are light years ahead of those currently available.

Microalgae produce far more oil compared with other bio crops – about 60 times more per acre. They take up less space, only require a little food and water and don't need insecticides or herbicides. You don't even need solvents to break down the algae's cellular structure and extract the oil; techniques have been developed to do it using soundwaves. Like other biofuels, algae use carbon dioxide to grow, so they're almost carbon-neutral. There's even been suggestions of growing algae farms next to power plants, using the CO_2 exhaust generated to feed the future fuel supply.

The biggest task is deciding which of the many strains of algae in the world can produce the best results. And this is where Mukhopadhyay's research takes things to the next level. Rather than rely on nature, the JBEI is modifying algae to breed new strains that can do exactly what we need. 'There's a pretty big bioengineering effort to develop microbial platforms to convert renewable carbon sources into the final target,' Mukhopadhyay says. 'Let's say you want a particular compound and there's an enzyme out there that makes it. You can stitch together a pathway, take a microorganism and give it the enzyme's capability.' Using bioengineering, scientists such as Mukhopadhyay can increase yields or ensure that the algae produce even better, more efficient fuels. Yet this research is just a snapshot of a wider picture that's still at an experimental level. 'We need to rethink the way we make decisions about fuels, and having options through science and technology is part of that,' Mukhopadhyay says. 'Biofuel research is about providing those options. It might not be a perfect solution but I hope that, in my lifetime, I will see biofuels playing a part in a big, very interconnected problem like climate change.'

Yet again, racing is already ahead of the game. While Formula E is pushing the boundaries of batteries, its exclusive rights to electric racing with the FIA means other formulas can't follow suit. Instead, series that are using ICEs have embraced second-generation biofuels. Formula One has used them since the advent of the hybrid engine era, with regulations requiring at least 5.75 per cent bio-components in its fuels. This increased to 10 per cent in 2022. In December 2020, the FIA even announced a fuel it claimed to be '100 per cent sustainable', refined exclusively from biowaste. While NASCAR isn't bound by FIA regulations, it's also chosen to focus on biofuels. Its three national touring series use a 15 per cent ethanol-based mix, which has been estimated to reduce the sport's carbon emissions by 20 per cent.

Biofuels aren't the only attempts at keeping ICEs alive: there's also pure chemistry. One of the biggest areas of focus is so-called 'e-fuels' – completely synthetic oils. 'Making hydrocarbons is pure chemistry, you don't have to just dig it out of the ground,' says Paddy Lowe. Now of Zero Petroleum, Lowe was previously with Williams, McLaren and Mercedes, and has seven world championships under his belt.* 'Petroleum has helped deliver most of the things in our society and it's weird to think that you can just make it from scratch ... what we've done in industrialisation is only the first half [of an energy cycle], we've done the engine and combustion part. The part we need to do now is delivering the [energy] material. And that means making petroleum fuels synthetically from CO_2 and water. It's exactly the same as a forest, only we're doing it in a factory.'

* Lowe was speaking at Autosport International Connect 2021.

Creating e-fuels isn't that complicated. First, you take water and split its molecules into hydrogen (H_2) and oxygen using electrolysis. Elsewhere, you have another process running, called the reverse water-gas shift reaction,* which takes carbon dioxide from the atmosphere and turns it into carbon monoxide (CO) and water. At this point, you then take the hydrogen and carbon monoxide and combine them together in a final series of reactions, run at about 150–300°C, called the Fischer-Tropsch process. This combines the two gases and condenses them into a liquid hydrocarbon. Ultimately, for about 1.5kg of water and 3kg of carbon dioxide, you end up with 1kg of hydrocarbons for fuel (along with about 3.5kg of oxygen, which you can also sell).

Porsche has already announced it will be building a plant for e-fuels in Chile, aiming to have it online in 2022 and producing 500 million litres of synthetic fuel by 2026. The catch, however, is that e-fuels take *a lot* of energy to produce – particularly when it comes to running hydrolysers to split water. The process is expensive, too, and incredibly inefficient: at best, only half of the potential energy is converted into fuel you can use for your car.

The reason Porsche chose to build their test factory in Chile is because of the country's immense potential for renewable energy. It's sunny and has sweeping gales coming off the ocean (the so-called 'Roaring Forties'), which allow Porsche to run its e-fuel refinery with solar power and wind turbines. It's a particularly interesting choice from a science

* RWGS reactions are going to be incredibly important in space exploration as they are a way of producing water. If we're going to travel to Mars, that's pretty essential.

history perspective, because the story of synthetic fuels has a close analogy with another landmark event that also began in Chile … and a war over bird poo.

★ ★ ★

About 5,000 miles to the south of San Francisco lies Arica, a small and impoverished town halfway down the west coast of South America. Here, the vast salty void of the Atacama Desert plunges precipitously into the deep blue of the Pacific. The Atacama is the driest place on Earth and rainfall is so unlikely many of Arica's households don't even bother with roofs – they just don't need them. This lack of precipitation also means that when an animal poops, their anal deposit doesn't get washed away. And so, over thousands of years of merry bowel movements, the local birds have turned the cliffs around Arica white with stinky, crystallised scat.

In the nineteenth century, these mountains were one of the most valuable resources on Earth. Bird poop, or guano, is rich in nitrates, which are fantastic for two things: fertilisers and gunpowder. Once this was discovered, it didn't take long for the entire world to exploit the resource, and wars were fought over anywhere that contained fossilised bird poo – such as around Arica. In 1879, mining taxes over these mounds of white gold sparked the War of the Pacific, with Bolivia and Peru on one side and Chile on the other. The war didn't go well for the Bolivian–Peruvian alliance. After the conflict ended, the borders were redrawn. Arica was originally part of Peru but today it flies a Chilean flag. Bolivia, meanwhile, lost its entire coastline to Chile and remains completely landlocked to this day. It's a mark of shame so acute that, every 23 March,

Bolivians celebrate the Day of the Sea, when children dress up like sailors and wish they could go to the beach.

But we haven't used Chilean guano for more than a hundred years. In 1909, a German scientist called Fritz Haber came up with an alternative. Nitrogen is a gas that exists in our atmosphere (in fact, *most* of the air we breathe is nitrogen) while hydrogen is easily obtained from natural gas. Using high temperatures, pressures and catalysts, Haber was able to pluck these atoms out and combine them to create ammonia (NH_3). The result, known as the Haber process, meant an endless supply of fertiliser for everyone. The Chilean bird crap barons were put out of business. Even so, the legacy of bird poo still lingers in science; in 1844, German chemist, Julius Bodo Unger, was investigating seagull faeces when he noticed a new substance, which he named guanine. Today, we know it's one of the four bases of DNA. Part of our genetic code is named after bird shit.

It's hard to imagine the modern world without the Haber process. It helped end famines and provided 'bread from the air'. The fertilisers made by Haber are often given as the reason the population on Earth soared from 1.7 billion to almost 8 billion in less than a century. Today, around 50 per cent of nitrogen in human tissue originates from the Haber process. It's a small wonder that Haber (and Carl Bosch, the colleague who helped him scale the process to mass production) won the Nobel Prize.*

---

* There is an incredibly dark twist to this story. Haber might have saved billions of lives with nitrogen fixation but he went on to develop one of the greatest evils in history. On 22 April 1915, during the First World War, Haber launched 'Operation Disinfection' against Allied troops outside Ypres, opening 5,730 canisters containing some 160 tonnes of chlorine gas. Haber had just invented chemical warfare.

The story of the Haber process shows how chemistry can solve these complicated problems and supplement natural resources. If we can do the same with hydrocarbons, then we'll be able to stop using fossil fuels for good. Even so, synthetic fuels are still in their early stages. And they also beg a simple question. As part of the process of making them, you need to split hydrogen from water. So why can't you just use hydrogen to fuel cars instead?

★ ★ ★

Say the word 'hydrogen' and most people think of one thing: the *Hindenburg*. On 6 May 1937, the German passenger airship caught fire as it attempted to dock in Manchester Township, New Jersey. Filled with hydrogen gas for buoyancy, when the zeppelin ignited it turned into a towering firework of bright yellow that caused 36 deaths and terrified onlookers around the world. It's a single piece of newsreel footage that soured the reputation of an otherwise exciting energy source.

Despite appearances, hydrogen – the lightest element in the universe – isn't very energy dense. While fans will point out 1kg of hydrogen gas has the same energy as 2.8kg of petrol, this is misleading: 1kg of hydrogen gas takes up about $11m^3$ – the carrying capacity of a transit van. 'Hydrogen is super-light, so to get any useable energy density, you've got to compress it a lot,' says Matthew Billing, a chemist at London South Bank University specialising in energy storage and hydrogen fuels. 'Most of the technologies are based on high-pressure cylinders at around seven-hundred bar [roughly 700 times the atmospheric pressure at sea level].'

Once compressed, however, hydrogen becomes viable as an option for electric motors. Hydrogen is taken into a fuel cell, where chemical energy is converted into electricity to power the car. 'There's several reasons hydrogen is great,' says Mark Jan Uijl from the Delft University of Technology in the Netherlands. 'If we want to make this world a greener place, then we have to figure out a way to store renewable energy for where the sun doesn't shine or wind doesn't blow. Hydrogen is a perfect way to do it. Right now, you can refuel a hydrogen car in about three minutes with as much range as a normal combustion car. So, it has all the advantages of current cars but also a green aspect to it.'

The level of environmental friendliness depends on where you get your hydrogen from. Currently, most hydrogen fuels are produced via gas or steam reforming – depending on the exact method, this is classed as either 'brown', 'grey' or 'blue' hydrogen. These are half-way options between fossil fuels and zero emissions. The alternative, which uses electrolysers to split water in the same way as e-fuels, is 'green hydrogen' – effectively carbon neutral if powered by renewable energy. 'The problem with green hydrogen is, economically, you need quite a big plant,' says Billing. 'There's just so many steps – electrolysis, storage, transportation – which can make things tricky.'

Green hydrogen is already attracting attention. In 2017, Saudi Arabia's Crown Prince Mohammed bin Salman announced his country would build a brand new 'smart city', called Neom, on the country's coast, stretching for 160km along the Red Sea. Costing around $500 billion, the plan is for 1 million people to live and work in a world with no cars but where everything is within a 20-minute commute thanks to advanced public transport. Perhaps most ambitious of all, the city also plans to have no carbon emissions by running on green hydrogen, with the electrolysis powered by solar and wind farms. And yet, as crazy as it sounds, making a city the size of Belgium run on hydrogen is easier than making it suitable for a road car. Unlike batteries, which have to be scaled *up* to be used in cars, hydrogen has the opposite problem: you actually have to scale it *down*.

One of the big problems is how you store the hydrogen. 'Hydrogen can permeate through most non-metallic materials,' says Billing. 'In the case of metals, it reacts with many of them. Materials selection aside, with a car there are extra complications, such as volume and mass constraints. BMW did a lot of work with liquefaction but the hydrogen has to be very, very cold to be liquified [about -253°C]. So now you've got to have something cold and high-pressure in your driveway. There are only about five stations in the world that support BMW's filling technology. Cars also get pretty hot, which could increase the pressure in the tank, which means you'd have to vent it. Imagine filling your car one day, not using it for a week and then finding your tank completely empty.'

Of course, you don't have to store hydrogen as H_2: with a little chemistry, you could attach hydrogen atoms to other molecules and break it off as you need it. Unfortunately, this limits the amount of hydrogen you can use. For Billing, the most promising options involve storing hydrogen as ammonia borane (NH_3BH_3).* 'It's a solid that you can easily mix with

* Ammonia borane is a good storing chemical because, despite looking like a hydrocarbon, it doesn't behave like one. The scientific explanation for this is that it contains boron, and boron is *super weird*.

other materials to make pellets. When you heat these up, the hydrogens click off. The hydrogen capacity of these materials can be as high as 15–20 per cent. The big problem is that the pellets are difficult to recycle.'

'Ultimately,' Billing concludes, 'making hydrogen practical in cars is a huge challenge and there are lots of areas that need improvement. It seems like every single piece of your car has to be specifically engineered in some way to accommodate for it.'

This is where racing makes a difference. At Delft, Uijl is the team manager of the university's Forze hydrogen electric racing team, a project led and run by students, many of whom have taken a gap year to work on the car full time. The Forze team's aim is to tackle some of the hardest aspects of viable hydrogen-powered cars and solve them. 'In the future, things will get smaller,' Uijl says. 'Maybe we'll be able to store hydrogen at higher pressure or as a liquid. But right now, the big challenge for us is space.'

Initially, the Forze team raced in Formula Zero, a series for vehicles with no emissions. They won the world championship in their first season but the team didn't want to stop there. 'We really wanted to show that these technologies can be applied in a way that normal people can use,' Uijl says. 'To do that, we realised we had to race against traditional, petrol-powered vehicles.' In 2017, the team did just that, with its Forze VII becoming the first hydrogen-powered car to compete against conventional ICEs in a sanctioned event. In August 2019, its successor, the Forze VIII, came second in the sports category of the Supercar Challenge in Assen. The podium shows that hydrogen-powered cars can do more than compete against their ICE rivals – they can beat them. The team's latest supercar, the Forze IX, can go from 0–60mph in less than three seconds and has a top speed of 186mph.

In addition to speed, the Forze team also emphasise that hydrogen, despite appearances, can be safe if handled correctly. 'So, indeed, hydrogen is flammable,' Uijl admits. 'But the safety of hydrogen vehicles is very high. The pressurised tanks have been tested. They've been shot at with guns, thrown from buildings. The only thing that penetrated them was an anti-tank gun! In a crash, the monocoque's survival cell would break before the tank. If there was a fire, it wouldn't affect the tank because of the high pressure. And, even if it did, hydrogen is lighter than air! The flames would go up and away, so the car wouldn't catch fire. It's a lot safer than combustion engines and fuels.'

The Forze team are pathfinders and the rest of motorsport is already following suit. In 2024, a hydrogen-power class will be added to the 24 Hours of Le Mans, and Red Bull Advanced Technologies, the technical arm of its Formula One team, has already announced an attempt at hydrogen-powered glory.

It's possible we might see Formula One cars running on hydrogen in the near future, too. Thanks to Formula E's exclusivity contract, Formula One can't go fully electric until 2039 at the earliest. Although it's likely to stay with biofuels for the next decade, it could be that the lightest element in the universe might be the perfect solution for the fastest sport on Earth.

For now, there's still work to be done with all of these potential fuels. The important thing to remember is that it's likely *none* of these options will be the sole method of fuelling our lifestyles in the future. 'Synthetic petroleum fuels aren't necessarily the best solution for personal transport,' says Lowe.

'Wherever you can electrify, that's the best thing you can do, as a direct supply will always be the most efficient. But creating synthetic petroleum [or biofuels, or hydrogen] isn't intended to replace that. We need *all* of these solutions for places where you simply cannot do electrification. Long range air travel is the classic one, but also things like combine harvesters. These machines are the reason we aren't all out working in fields, cutting corn – they are intensely powerful and are running long shifts and are already on the weight limit for what could be powered by diesel.'

Simply put, electric vehicles are still the best answer ... unless you can't use an electric vehicle.

And when it comes to sustainability, there's an even more pressing shortage that racing has to address. For more than a century, an invisible crisis has constantly threatened one of the most essential materials on the planet. Now, this impending disaster has come to a head and needs drastic action.

The world is running out of rubber – and it's hard to go racing without tyres.

CHAPTER FIFTEEN

The Terrible History of Tyres

The Amazon river basin of Brazil. A land of muddy waters rich in chocolate tones, of vegetation poking itself free from the flow, of eddies filled with the telltale bubbles of aquatic life. I've set out by ship to pierce its depths; with no roads and few flights, it's the only way to reach my destination. At times, the river is narrow enough that it feels much like any other; a dirty, rushing torrent that locals use as a highway, cutting past stilted houses in long, slim motorised canoes. We pass so close to the tree line that monkeys can be seen dancing amid the branches of a thick, verdant rain forest that stretches to a jagged edge on the horizon. Then, with a twist of a river bend, the ship arrives in a confluence. Now the waters stretch on either side of us as far as the eye can see, thick channels as smooth as a mill pond, separated by thin slivers of low-lying islets. It's astonishing to know that you're 250 miles inland, yet all you can see is water.

This incredible world explodes into colour as we sail beyond trade stations with tall antennas, disused mining jetties and old ferryboats converted into floating houses safe from animal attack. On the second day, we pass remote Itacoatiara, its church gleaming white in the equatorial sun. Further along the river, ruddy-brown wisps billow in the sky, tails that lead to the latest logging camps and deforestation. In this land we plunder the Earth for our own comfort.

The next day, the ship arrives in Manaus, a two-million-strong metropolis in the heart of Amazonas, 800 miles from the coast and reachable only by boat or plane. Times here are to be witnessed with all five senses. It is a world of sun-soaked humidity, alive with samba, colours, music and fabrics, where women strut for companionship and old men, skin like burned mahogany, offer 'Água! Água! Água!' Hawkers trade on the wharf, selling rows of stuffed piranha caught and glazed for tourists. Pink dolphins dance along the riverbank, begging for scraps from fishermen. Candles to the saints gutter and flutter in the empty cathedral. Young girls dance in open courtyards, gripping the hems of their coloured dresses and twirling round in wild spiral shapes as they practise the carimbó. It's a place where adventure crackles in the electric air, scented with citric zest and rich in the taste of tropical mystery.

I could follow the world's mightiest river onwards into Peru, but my course lies elsewhere. Outside the city is the 'meeting of the waters', where the Rio Negro – its clear water seemingly black from the thick mud of its riverbed – flows into the cloudy Amazon. The two currents refuse to mingle for several miles, racing together before gradually losing their sharpness until, like milk in coffee, they blend and swirl. An hour after arriving in Manaus' bustling docks, I'm on a speedboat cutting north along the Negro, past the longhouses of Amazonian tribes with white flags fluttering outside to indicate they're accepting visitors. Finally, amid lashing rain, I arrive at a small plantation. Outside a humble wooden shack are a series of charred, foot-long bales, stacked like alien seed pods. Beyond, neat rows of trees spread out into the undergrowth, their arrow-straight trucks criss-crossed with thousands of vicious-looking scars.

This is the reason I've come. Rubber.

The trees are *Hevea brasiliensis*, the para rubber tree. The plantation itself is a working museum, originally constructed as a set for the film *A Selva* (The Jungle), on the condition it would be left to educate visitors about the rubber boom. It's an exact re-creation of thousands of such settlements in the mid-nineteenth century; all that's missing are the slaves. In the Victorian era, those toiling at such a plantation would have been some of the 4 million Africans shipped to Brazil in chains, or indigenous people enslaved to work the harvest. The conditions were terrible, including the rape and mutilation of women and children to force the men to work. Such abuses continued until Brazil finally abolished the slave trade in 1888. Rubber is a material with its history soaked in tears.

One of the plantation's modern workers beckons me over with a long blade, a cross between a machete and a bill hook. Taking his weapon, he swings down in a deft slash to rip the flesh from a nearby tree. A pure white river bleeds from the wound. It's fresh latex, a sticky goo the tree has evolved to defend itself against insects. As the latex weeps from the edges of the gash, the worker pins a small metal cup to the tree to collect the trickling flow. The cut produces little more than a dribble but, with careful management, a tree can be milked like this thousands of times over its lifespan. 'Work was hard,' my new friend says. 'The slaves had to do it during the night, it was too hot during the day. They would work by torchlight from burning pig tallow.' The heat had nothing to do with the comfort of the slaves; in low temperatures, latex concentrates at the bottom of the tree, so working at night produced higher yields. 'Each worker would tap three hundred trees a night. They were all expected to produce sixty kilograms of latex a week. And it was dangerous. The Rio Negro doesn't have mosquitoes, there's no malaria. But there are jaguars.' Once gathered, the latex is pressed into sheets to remove water, hung in smokehouses to dry and then bundled up for shipping.

It's a horrifying thought that this toil – slaves breaking their backs in the heart of the world's largest rainforest, working at night with death lurking in every shadow – gave us rubber bands, latex gloves and the soles of our shoes. You'll find rubber in our clothes' elastic fibres and as shock absorbers on the edges of personal electronics. You'll even uncover it amid the foundations of skyscrapers built in earthquake zones, such as Taipei 101. But today the most common use, about 70 per cent of the world's natural rubber reserve, is making tyres.* And it's worth looking at the material more closely to understand why it's so important.

★ ★ ★

In a rubber tree, latex is produced inside cells called laticifers. These secrete a messy goo of proteins, fatty acids and resins, along with long stretchy chains of the polymer *cis*-1,4-polyisoprene. These chains are normally screwed up in a mess but, if you pull dried latex material, the chains will stretch out, like tugging a line of spaghetti from a bowl of pasta. Natural rubber polymers have a glass transition temperature (when they go all crystalline and snap) of about -70°C.† This means that when you pull them – within reason – they'll happily slip back into their normal pasta bowl state if you let go.

* The world's leading manufacturer of tyres isn't Bridgestone, Michelin or Continental; it's Lego. Around half of Lego sets contain wheels and the toy company makes around 700 million tyres a year to meet demand.

† Different rubber compounds have different glass transition temperatures. Perhaps the best known example of this is the *Challenger* space shuttle disaster, where an O-ring seal failure occurred as the material lost its resilience at just 0°C.

Natural rubber latex formed into products can stretch to 10 times its length. It is waterproof and it doesn't conduct electricity. It has knotty tear resistance, which means it's terrific under deformation and dampens vibrations. It can be made into thick lumps or it can be woven into a micro-thin skin. It is, quite simply, a brilliant material. It just needs to be treated right.

Rubber was introduced to the US in the early 1830s, but was a commercial disaster. It looked like chewing gum and in hot weather it melted into a wretched, stinky glue. In 1834, after being inundated with returns during a heatwave, the directors of the Roxbury India Rubber Company had to bury $20,000 of their melted stock in a pit. Soon after, an inventor, Charles Goodyear, decided to solve the problem of gooey rubber. Fresh out of debtor's prison, Goodyear set his family up in a New York tenement bedroom, where he worked trying various powders to dry out the rubber's stickiness. After numerous experiments – including a dead end where he melted 150 mailbags for the US government – he was destitute. The Goodyears moved often and eventually found themselves in Woburn, Massachusetts, where local farmers gave their children milk and potatoes to survive. In the winter of 1839, Goodyear was demonstrating his latest attempt to dry out rubber at the town store, when a group of onlookers started laughing at him. Angrily waving his new sulphur-and-rubber mixture at them, Goodyear slipped. The sample flew from his hand and fell on to a nearby stove. To Goodyear's amazement, the rubber didn't melt. Instead, it turned hard.*

---

* This is just one of multiple different legends about how Goodyear's rubber ended up on a stove; all that's truly certain is that it was an accidental discovery.

In a complete accident, Goodyear had discovered that, by heating rubber latex with sulphur, an interesting chemical reaction occurs. At high temperatures, sulphur bonds with *cis*-1,4-polyisoprene, forming short chemical bridges between adjacent polymer strands. These bridges limit how far each strand can move, increasing the rubber's durability and turning the material weatherproof. Goodyear called the process vulcanisation, after the Roman god of fire.*

Vulcanisation was a game changer. By altering the length of the sulphur chains, chemists could control the material's heat resistance and rigidity, all while thermosetting the rubber into whatever shape they required. Soon, synthetic rubbers emerged, too. Changes to the synthetic/natural mix, the vulcanisation process and the mould can be used to create millions of different products and thousands of different tyre designs.

Racing tyres aren't the same as the ones on your car. They're wider, with a more rigid internal structure and are designed to withstand far greater *g* forces when breaking. They're also far more expensive: a set of Formula One tyres costs about £2,000. In dry conditions, slick tyres are used – a flat surface that maximises contact with the road. Initially, their smooth surface means they have very little grip, which is why drivers have a formation lap to warm their tyres, using friction with the road surface to generate heat and cause the rubber's long

* Goodyear and his descendants have absolutely nothing to do with the billion-dollar Goodyear Tire & Rubber Company. In fact, money woes continued to plague the inventor and he died, $200,000 in the red, in 1860. Perhaps the moment that best sums up his life occurred in 1855, when Napoleon III of France awarded him the Cross of the Legion of Honour for discovering vulcanisation. True to character, Goodyear was in a Parisian debtor's prison for unpaid hotel bills.

polymer chains to stretch out. This makes the tyres more viscous (or sticky), allowing them to deform as they touch the road and hug any imperfections in the surface. The result is more grip and traction.

Over the course of a race, an F1 car's tyres will heat to about 120°C. But there's a sweet spot for a tyre's temperature. Too hot on the inner part of the tyre and it will cause the gas inside to expand and tyre to burst; too hot on the outside and it will cause clumps of the tyre to be stripped away. And, as a tyre wears, hot and sticky rubber will begin to flake off and stick to the outer layer, breaking up the surface contact and reducing grip. When that happens, it's time for a pit stop. In Formula One, there are five different slick tyres, all with different degrees of softness, and the tyre manufacturer chooses which three compounds will be available before the race. The softer your tyre, the better your grip, although it will also degrade earlier in a race. And if Formula One's options seem a bit excessive, keep in mind a NASCAR team – whose objective is to literally *go around in an oval* – has 18 different sets of tyres to choose from over the course of a season. Formula E, meanwhile, is a little better; it already uses the same tyres in both wet and dry conditions, and limits drivers to a single set of tyres for the race weekend (plus two spares, carried over from the previous race.)

Ah yes, wet conditions. Racing doesn't just happen in the dry, and standing water on a track will form a liquid barrier between the tyre and the road. At best, this means a slower lap time; at worst, it means you aquaplane, skid off and crash. In Formula One, there are two tyres that are treaded: the intermediate and wet tyres. These treads are essentially grooves designed to displace water, shoving it out of the way so the rubber meets the road and you don't go spinning sideways. Wet tyres (and, for regular cars, winter tyres) are typically made of softer rubber, too, which helps to maximise this grip. These treaded tyres are closer in design to the ones you'll find on a road car but they're still far more advanced. A Formula One wet tyre can displace about 60 litres of water a second – about the same as having a bath – compared with about 13 litres a second for a tyre on a standard road car. But, then again, your mum isn't racing around Monza at 150mph when she pops out to do the weekly shop, so she probably doesn't need that level of performance.*

As you can see, rubber is pretty important, both for racing and for our daily lives. The problem is that we've already reached the limits of the material's sustainability. And the reason for this impending disaster lies in yet more truly awful moments of history.

★ ★ ★

It didn't take long before other countries wanted a piece of Brazil's rubber trade, with predictably violent results. In 1884, German chancellor Otto von Bismarck called 14 colonial powers together to formalise carving up Africa among themselves. After dividing up the continent at the Berlin Conference, the powers discovered they'd left a large gap. Rather than grant the land to, you know, the people who

* This is massively abridged version of what's happening and how tyres are constructed. For more details, see *Sticky* by Laurie Winkless, who has an entire chapter dedicated to how Formula One tyres operate.

lived there, Leopold II, King of the Belgians, came up with an alternative solution. He suggested the land was given to him – not to Belgium, but as his own private fiefdom. In return, he promised he'd look after the territory as an act of philanthropy. The powers agreed and the Congo Free State was created, run as a private corporation with Leopold the sole shareholder.

Unfortunately, Leopold was a callous dickbag who'd been lying through his teeth. He'd already been working for years to carve out a personal empire in Africa, even hiring famed explorer, Henry Morton Stanley, as his agent. Leopold had tricked the world into granting him a 2.3 million km^2 private estate that he could exploit without mercy. By 1891, he had dropped any pretence of altruism and turned the Congo into the world's largest forced labour camp. Locals were soon ordered to work on giant plantations of a rubber-producing vine, *Landolphia owariensis*, to compete with the Brazilians. Rather than collect latex in cups, workers were instead forced to smear it on to their skin, dry it in the sun, then rip it off for collection. Leopold didn't care if their skin and hair came off with it.

Anyone who refused to work was whipped, killed or had their villages burned to the ground. Workers who failed to meet their rubber quota were executed. As proof of death, Leopold's colonial militia would collect the hands of their victims. As a consequence, if a village thought it was going to miss its quota, it would attack a neighbouring settlement, mutilating innocent people so the rubber tally could be supplemented with baskets of severed limbs. The 'red rubber' trade only ended in 1908, when Leopold was stripped of power. Unbelievably, he got away with his atrocities and died peacefully the next year. By then, an estimated 10 million Congolese had been murdered.*

While the horrors of the Congo Free State were being uncovered, the British were already doing what they do best: nicking stuff from other countries. In 1876, explorer and bio-pirate Henry Wickham visited plantations along the Amazon and helped himself to rubber tree seeds. In the course of a year, he managed to gather around 70,000 *Hevea* seeds from plantations near Santarém before smuggling them back to Kew Gardens in London as 'academic specimens'. The expert gardeners at Kew managed to germinate 2,700 of Wickham's pilferings, which were promptly dispatched to British colonies with tropical climates that suited the tree. In decades, rubber plantations sprung up in Sri Lanka, Malaysia and Indonesia. Rather than farming natural groves, these were purpose-built plantations, mostly owned by small sharecroppers. In Brazil, rubber stocks were always limited by South American leaf blight; in Southeast Asia, there was nothing to stop the new rubber barons. Soon, the British outpaced the Brazilians with their own tree.

The Brazilians tried to fight back. In 1928, they agreed to let Henry Ford build a new town in the Amazon on the river Tapajós, 190 miles upstream from Santarém. The settlement, Fordlândia, was intended to dodge the British monopoly and provide the Ford Motor Company with its own, exclusive rubber supply. It had American-style housing, along with a

* We need to be mindful that history could repeat itself. Earlier in the book, I mentioned how cobalt is used in batteries. The number one supplier of cobalt in the world, responsible for 65 per cent of production, is the Democratic Republic of Congo. Already, men, women and children are being exploited there to get the metal.

hospital, school, hotel and even a golf course. Unfortunately, the town was a complete disaster. While Ford's managers knew about cars, they knew nothing about agriculture. Trees were planted on hilly land where machinery couldn't be used, and the crop was grown too close together, resulting in attacks by ants and other jungle critters, as well as problems with leaf blight. Many workers succumbed to yellow fever and malaria, and discontent spread thanks to Ford's prohibition of alcohol, women, tobacco and (worst of all) games of football. In 1930, sick of being made to eat hamburgers in the cafeteria, the workers revolted and took over, chasing the Ford managers into the jungle. The Brazilian army was called in to quell the uprising but even that couldn't save the project. Fordlândia was abandoned in 1934. When a second settlement also failed, Ford pulled out of Brazil. While *Ford Vs Forest* isn't as cinematic as *Ford Vs Ferrari*, the company still lost the equivalent of almost $300 million.*

In Asia, the British-led rubber trade continued to boom. Today, around 14 million tonnes of natural rubber are produced a year and the world's leading exporter is Thailand. But while there's no question Wickham's theft was the most successful heist in history – one that stole an entire industry worth billions of dollars a year – it also had doom written into its very DNA. 'Not only is the industry all geographically centred around one very small part of the world,' explains Katrina Cornish, a professor of bio-emergent materials at The Ohio State University. 'But the base genetic diversity is extremely

*The revolt over Fordlândia's hamburgers in 1930 is known as the *Quebra-Panelas* (breaking pans). The second settlement, at Belterra, didn't have any uprisings but fizzled out as it wasn't cost-effective.

low. All the trees are clones. For miles and miles, you have genetically identical trees. And that means if one catches a disease, they all go down.'

In 2019, that's exactly what happened. Two leaf blights, *Pestalotiopsis* and *Neofusicoccum*, hit Asia. 'By the end of the year, we had lost a million tonnes of production,' Cornish says. Through 2020, researchers estimate production was down 1.4 million tonnes, although it's hard to give a precise number because of the disruption caused by Covid-19. But even if the blights are stopped from damaging more trees, 10 per cent of the world's natural rubber stock has already been devastated. Farms unable to work because of the Covid-19 pandemic, extreme weather events and another two leaf blights mean the knock-on effects could last decades.

But the demand for rubber isn't going to go away. In fact, it will likely increase to 30 million tonnes a year in the near future. For modern life to continue, we need an alternative to the rubber tree. There are three ways we can meet this gap: recycling, synthetics or natural alternatives.

Let's start with recycling first. NASCAR recycles around 300,000 tyres a year, while Formula One scraps around 1,800 tyres per race (including 560 wets that might never be used), which brings their tyre consumption to about 36,000 a year. The problem is that recycled vulcanised rubber is never as good as a fresh source and is only used as an additive. At the moment, waste rubber is ground down and used either for low-grade materials (mats and flooring, or mulch for children's playgrounds) or blended with wood, coal and biomass and simply burnt for energy; Formula One's tyres, for example, are used to power a cement factory near Didcot.

At best, only a tiny portion of the world's rubber demand can be met by recycling. 'Most recycling is able to take just a small fraction of already-used rubber,' Cornish says. 'Perhaps 10 per cent of a tyre's filler can be made up of recycled material. Chemists are looking at biodegradable, crosslinking systems, where you could take a finished product and unvulcanise it. That would need something that doesn't use a sulphur cross-linking system because sulphur bonds are hard to break. We are currently looking at chemistry that could permit more recycling.' We're not there yet, though. At best, recycling might be able to make up 10–15 per cent of the demand for rubber. And that's it.

Chemistry brings us on to the next option: synthetic rubbers, created and designed in the lab. These have been more successful; part of the reason for Ford's failure in the Amazon was the advent of synthetic rubber. By the mid-twentieth century, cunning chemists were able to make a host of different rubber substitutes. This slashed the price of natural rubber (less of it was needed) and made any cultivation with high overheads unsustainable. Yet while synthetic rubber is fantastic for some properties, such as solvent resistance, no lab in the world has been able to make chains as long – or as neat and ordered – as we can find in nature.

In chemistry, otherwise identical molecules can be arranged in different geometric shapes. This is called *cis–trans* isomerism and can give a molecule very different properties. In natural rubber, virtually all the molecules in a polyisoprene chain are in a *cis* configuration, meaning all of the functional groups (the bits that react with stuff) are on the same side. But when you try and make it in the lab, you get too many in a *trans* configuration where some functional groups are on the wrong side of the chain. 'I always compare it with football stadiums,' says Carla Recker, a materials chemist for Continental who describes her job as trying to understand the 'black magic of rubber'. 'If you're making a synthetic polymer, you're creating a small football stadium filled with twenty thousand people. And, if you asked them to connect in a long chain, so everyone's left hand is holding their neighbour's right, about five hundred of the twenty thousand will do it wrong – maybe cross their arms and use their right hand, or something stupid like that. That's what we can produce. In nature, there are 100 thousand people in the stadium and only fifteen of them hold hands the wrong way. And that makes a difference, especially when it comes to a material's endurance.'

Put simply, nothing beats the real thing. 'If you need heat dispersion – which is important in tyres – and strength combined with deformation,' Cornish says, 'natural rubber cannot be surpassed by anything synthetic. It's also the best thin film material, which is extremely important for personal protective gloves.' And, while synthetics can be made in greater quantities if needed, the supply chain isn't in place to meet unexpected demand. The Covid-19 pandemic, for example, doubled the demand for natural and synthetic gloves and the world needed to make 20,000 gloves a second to limit contagion spread; by the end of 2020, nitrile latex was sold out through 2022. The world had simply run out.

That leaves the final option: natural alternatives. 'Natural rubber is made in about two-thousand-five-hundred different species,' Cornish explains. 'Some of it is made in individual cells, some as latex. The amounts vary, the chain length varies, and the composition of the membrane – which becomes part of your ingredient – varies.' Some have more protein and less lipid, others the reverse. 'If you look at something like *Ficus elastica* [a common ornamental house plant], this has big rigid particles with lots of wax – you can crack them like an egg shell and the rubber centres pour out. This type of rubber is not good for glove films.'

From these 2,500 rubber producers, there are two viable alternatives to the para rubber tree: *Parthenium argentatum*, or the guayule shrub, which grows native to the Chihuahuan Desert of Mexico; and *Taraxacum kok-saghyz*, or the rubber dandelion, originally from the steppe of Kazakhstan. Both of these plants have been investigated as alternatives to rubber trees before, first in the 1920s, when leaf blights hit South America and Asia, and then during the Second World War.

The Allies, whose rubber supplies were hit by the Japanese occupation of Malaysia, soon turned back to Brazilian rubber, an industry that had already collapsed. Rather than give up, Brazil drafted an army of 55,000 *soldados da borracha* (rubber soldiers), often kidnapping young men off the street, and sent them into the Amazon to harvest the crop for the US. It's estimated that half of these rubber soldiers died in the jungle, killed by animal attacks or disease. In California, the US also grew 35,000 acres of guayule at the Manzanar prison camp, where Japanese-Americans who had been interned were made to work in the fields. Nazi Germany, meanwhile, began by growing rubber dandelion in Raisko, a sub-camp of Auschwitz. There, 320 female prisoners, many of whom had degrees in chemistry, biology and agriculture, experienced horrific torture and abuse as they tended to the crop. Elsewhere in the Auschwitz complex, other scientists, mostly Jews such as the chemist Primo

Levi, were forced to work on synthetic alternatives. The Raisko project lasted until January 1945, when the women were forced on a death march to other concentration camps.

Once again, rubber is linked to the worst moments in our past.

★ ★ ★

Today, there is a resurgence of interest in guayule and rubber dandelion. Both plants have attracted interest from tyre companies, with Bridgestone working on guayule in the Arizona desert, and Continental and Linglong heading rubber dandelion efforts in Germany and China, respectively. Guayule's main advantage compared with rubber trees is its robustness, allowing it to be grown in harsh environments with very little management. The shrub's rubber also lacks the ingredients in traditional rubber that cause latex allergies and it can make much thinner materials than para rubber, while retaining the same strength. It's Cornish's favoured option. 'It's not very branched – it has a linear polymer that's soft and stretchy,' she says. 'You can make thinner films, absolutely wonderful condoms, or other products such as weather balloons that carry heavier loads.'

About 5 per cent of a guayule shrub contains rubber latex. While it can't be tapped like a rubber tree, it's a perennial crop with a harvest cycle as short as one year after the first two years of growth. And the remainder, or bagasse, is high in terpene resins. This means it can be burnt as a biofuel to cover the costs of extraction. Better still, as guayule is grown in arid environments, there's a vast amount of open, uncultivated land just waiting to be used. 'In the Second World War, the emergency rubber project assessed how much eligible land there was in the US,' Cornish says. 'They came up with 123 million acres. In Arizona alone, there are eight million acres of land, right now, where you could grow guayule, most currently uncultivated.'

Over in Europe, Recker's interest is *Taraxacum*. In 2007, her team was approached by German research organisation the Fraunhofer Society to look at one of the best options to replace the world's rubber stocks. 'They were studying rubber dandelions, looking at how natural rubber was formed in the plant. They approached us with something the size of a piece of chewing gum and asked us if it was the kind of rubber that could be used to build tyres.'

It turns out it was. Dandelion's latex – the same stuff you can squeeze out of its stalk as white milk – is very similar to a rubber tree's. 'As kids, we used to call it *kuhblume*, or cow flower!' Recker laughs. 'But most of the latex is actually under the earth – it's more or less completely in the root of the plant, so you have to grow and harvest it.' Once done, you essentially grind it up to create the chemistry equivalent of a dandelion root milkshake then separate the rubber out. 'And once you do that, you end up with a natural rubber you can use for tyres. There's no real difference.'

A rubber dandelion isn't the same as its garden variety counterpart: it's hardier and has a much longer root system, which means there's more rubber to be harvested. On a chemical level, the advantage of *Taraxacum* over guayule is that it's more suited to rubber uses with heavy wear – such as tyres. And, while guayule could dominate deserts, *Taraxacum* is more likely to be found in temperate regions. 'If you were to replace the full natural rubber consumption of the world with dandelions, you'd have to have a space the size of Switzerland and Austria combined,' Recker says. 'It sounds a lot, but spread it over the world. It's not like you'll see dandelions everywhere.' And, as with guayule, *Taraxacum* can be used as a biofuel, too.

It's likely that the world's future rubber crop will be a combination of these options. It could be that Europe grows its rubber from dandelion, semi-arid countries from guayule and the tropics from *Hevea* plantations. But that's good for everyone – it cuts down on transportation costs (both financially and in terms of carbon footprint) and makes the world's rubber supplies more sustainable. Unfortunately, this change won't happen overnight. 'It's got to grow, right?' Recker says. 'You can't just instantly have one thousand hectares because you don't have the seeds in place, or machines to do the harvest. And weed control isn't as easy in your home garden, you can't just do it with a fork! The challenge is creating the value chain and these things take time to become established. The last crop humans industrialised was sugar beet; that was 150 years ago, and [the process] is *still* being refined. We are doing things for the first time – there's no one we can ask for help. It's going to take time.'

Given the suffering that's occurred in the name of rubber, we owe it to the world to ensure that we get it right this time. And the good news is that, already, Continental has found a perfect partner in motorsport to test the material's capabilities. It's a new racing series, emerging from Formula E and backed by motorsport's biggest stars.

Only this time, the racing isn't happening on track. The lights are going green at the edge of the world.

CHAPTER SIXTEEN

Going to Extremes

Ushuaia, Argentina, is about as far from a racing track as you can get. It's about as far from *anywhere* as you can get. The southernmost city in the world, Ushuaia is nestled among snow-tipped peaks at the furthest reaches of Patagonia. This is the end of the road for those who brave the Pan-American Highway and the start of an adventure for those pushing on by sea to Antarctica. I've arrived by the same ship that took me down the Amazon, slipping through the chilly air of the Beagle Channel, past colonies of inquisitive Magellanic penguins and groups of seals who think the ship's bulbous bow is the perfect spot for basking.

To the north, across the barren inhospitability of Tierra del Fuego, the Straits of Magellan await those who venture past the Cape of 11,000 Virgins, pushing on to Punta Arenas and the resonant blue-hued glaciers of southern Chile. To the south, there's only a clasp of houses at Puerto Williams and the lonely lighthouse that rises out of the misty, perilous rocks at Cape Horn.* I am truly at the ends of the Earth.

I'm here because it's one of the opening rounds of the latest venture into the world of electric motorsport: Extreme E.† Founded by Formula E's supremo Alejandro Agag, it's an FIA-sanctioned, off-road racing series using electric,

---

* I visited Cape Horn shortly before my ship docked in Ushuaia. The lighthouse keeper lives there permanently with his wife and three children, the youngest of whom is a babe born at the station. Theirs is a fog-shrouded, ghostly vigil.

† At least it was supposed to be; the race was cancelled after this book was submitted due to the Covid-19 pandemic.

custom-built SUVs. They're coming to remind everyone that the glaciers and permafrost I can see before my eyes might be gone in two decades if we don't act now.

The principle of Extreme E is to showcase sustainability at the far reaches of the globe. Rather than race on track, the teams face off on areas affected by global warming, aiming to show people comfortable in their cities the embers of a fire already sweeping our world. It's the only sport I know with its own climate change scientific committee.

Extreme E has already attracted heavy hitters. Formula One world champions Sir Lewis Hamilton, Nico Rosberg and Jensen Button have all entered a team, as have rally superstar Carlos Sainz and American IndyCar teams Andretti Autosport and Chip Ganassi Racing. Uniquely, these teams have gender parity from the start: one male and one female driver to ensure equality. It bills itself as an electric odyssey: part racing series, part adventure across the planet. Everything the teams need, from the cars to the pit equipment, is housed on the sport's floating base, RMS *St Helena*, a former Royal Mail cargo-passenger ship, which will zig-zag across the globe and reduce the freight carbon footprint by 70 per cent. I've already seen where one of the races will be held – Santarém, downriver from Manaus, where Ford tried to build a factory and Wickham stole his rubber seeds.* Other locations include Senegal and Greenland. Wherever it goes, Extreme E wants to maintain a legacy of sustainability for the local communities, which are among the most affected by climate change.

'Formula E came in a moment we felt there was a gap in motorsport,' Agag tells me across a video link. 'There was a gap

---

* Guess what? This race was also cancelled after the book was submitted due to the Covid-19 pandemic.

that was evolving, but sport wasn't evolving in the same way. Extreme E is just the logical extra step. It's challenging, but it's going to be incredible. Racing in the Arctic Circle will be incredible! Racing in the Amazon rainforest will be incredible! Having everything on just one ship will be incredible!'

Agag is a mercurial character. He grew up in Madrid, Paris and New York, becoming a banker and Member of the European Parliament, as well as the owner of the Addax GP2 racing team and Queens Park Rangers football club. He is a man who straddles many worlds and knows many things. And, if he doesn't have the answer, he can always call someone who does.

As much as Extreme E is a showcase, it's also a test platform, he says. 'The technology challenge is very relevant for us. At extreme temperatures and surfaces, you need to develop powertrains in different directions [than in cities]. There's not much evidence about the capability of electric cars off road and we need to explore that. In Formula E, we focus on efficiency, range, battery power, software and so on. But in Extreme E, we are working on the battery's resistance to vibrations and temperatures. These are lessons that can be applied to all parts of the world.'

There's a lot of scientific data that can be picked up from the series' custom-designed car, the Odyssey 21. Built by Spark in Paris, France, it's capable of going from 0–62mph in 4.5 seconds and tackling gradients of up to 50 degrees. It's also designed for sustainability. The suspension, brakes and uprights are all symmetrical, so parts from the front left could be mounted on the rear right. That means fewer spares need to be taken. The framework is made from niobium, a ductile metal with a similar hardness to titanium, coated in Bcomp's flax panels. Even its tyres are unique, designed by Continental to tackle the problem of a single off-road tread that can handle everything from grit and gravel to snow and sand. They're made from Carla Recker's dandelions, of course.

Perhaps the biggest leap forwards is in the E-SUV's 400kW battery. As discussed earlier, batteries operate in a temperature sweet spot. But if the technologies discussed in this book are ever going to succeed, they need to be able to go anywhere, from the tropics to the polar regions, and in places without good road infrastructure or reliable electricity grids. When you're 500 miles from comfort you need the best solution possible.

It's something Williams Advanced Engineering, as Extreme E's battery supplier, has been working on. 'When we were first contacted about this project we thought it was extraordinary and crazy,' says Williams Advanced Engineering principal engineer Glen Pascoe.* 'But then we stopped to think about how it can be done.' Williams supplied the battery for Formula E's first four seasons, and its engineers became expert in assessing the health of batteries and minimising variations through impact (something that's pretty important when you go off road). 'The key challenges [with Extreme E] are environmental,' Pascoe says. 'Ambient temperatures can range from -30°C to 40°C, so how do you manage cell temperatures? How can we maintain electrical isolation and heat exchange in a place with 100 per cent humidity? And any time you've got a risk from foreign media, such as mud, sand or dust – which are all present – you're risking the safe operation of a battery. And you've got to service these engines in adverse

* I caught up with Pascoe at the Professional Motorsport World expo at Cologne in 2019.

conditions with no backup facilities. We're entering into the unknown.'

Williams achieves this with software to continually measure and optimise the battery. In particular, this means focusing on coolant. As discussed earlier, methods such as graphene coolants are being investigated by F1 teams. But Williams has gone a different route. 'We use ambient heat exchangers and very high thermal conductivity components in our packs,' Pascoe says. 'When a battery is recharged, it generates heat. But we've been able to create a storage system that can [dissipate] heat in ambient conditions with only the energy consumption of a simple fan. It's still not going to work in all of the locations within the timescales required [for Extreme E] – we can't do it in the Sahara Desert, for example. But, by optimising heat transfer throughout the pack, we can have a lower energy impact.'

The feedback from Extreme E will drive the next generation of battery development, making our road cars tougher and more resilient. But they're not the only racers looking to push the boundaries. Back in Monaco, I managed to grab some time with another of Venturi's mechanical experts. Franck Baldet is the team's chief technology officer. A mechanical engineer, he initially worked for Ferrari for its innovation department, looking at how to take Formula One technology to its road cars, before he realised electric vehicles were the future and moved to Venturi in 2010. Since then, he's worked on a series of 'global challenges' to see just what is possible. 'We needed to show you can use electric vehicles anywhere,' Baldet says. 'The first challenge we did was from Shanghai to Paris. The driver, Xavier Chevrin, and his partner, Géraldine Gabin, had to do it alone and recharge the car every day. The challenge was basically, "Where do I plug in my electric car?" and he was knocking on doors in Kazakhstan asking if he could use their electricity. Then he drove from Kilimanjaro in Kenya to Okavango in Botswana. Everyone thought we were completely crazy, but the aim was to demonstrate that electric cars are a reality and can work anywhere. And now the challenge is how to evolve and improve the technology.'

It's this drive that led to the team uniting with The Ohio States University's Buckeye Bullet, aiming to become the first electric vehicle past 400mph. But it's far from the only project they're looking at. Already, Venturi has developed an all-electric polar exploration vehicle. In 2019, it travelled 42km across British Columbia in Canada at temperatures as low as –30°C. Its next destination is Antarctica – and it wouldn't be possible without the skills the team picked up through racing.

'What we learn on one side of the business, we can use on the other,' Baldet says. 'They're not identical parts but the ideas and concepts are the same, from Formula E to Antarctica. We take things from one project, apply it to another. It all crosses together. For example, efficiency is very important in Antarctica – once you're there, you don't have any charging stations! So Antarctic vehicles need to be able to optimise energy use, in the same way as a car on track.'

Thanks to the incredible amounts of data generated by Formula E, the team also knows exactly how much heat is generated by their parts. 'In Antarctica, you have to fight against the cold,' Baldet continues. 'So, we could pick up the heat generated by an electric motor or inverter – because there's always some energy loss through heat from the powertrain – and use that to heat the cockpit, or we can take the heat and use it in the battery. It's all optimisation and all work that we got from our team in Formula E.'

And, if you think this is only useful for Antarctic explorers and will never touch your life, think again. In 2018, Jaguar Land Rover launched a race series for its I-Pace SUV, collecting data from the car on the track and looking at how to apply it to road cars. The manufacturer used the information to improve regenerative braking and battery performance, as well as temperature control and even when to close radiator vanes to improve aerodynamics. In 2020, the company sent all of this data to *every I-Pace in the world* as a free software update, increasing their distance by about 5 per cent without the car even leaving its owner's driveway.

We live in a world where racing breakthroughs no longer trickle slowly into road cars. They download instantly without us even noticing.

★ ★ ★

Back in Ushuaia. Back at very end of the Earth. It's here, among the rough rocks and cold winds, that everything covered in this book will be applied so it can then be passed on to you. Extreme E uses electric batteries recharged by on-site hydrogen stations; it uses 3D-printed advanced metals in its chassis, coated with natural fibres, on wheels made of sustainable rubber. The aerodynamics come from countless simulations and CFD. Safety for the races is provided by John Trigell and MDD Motorsport Medical. Audiences will be able to watch in virtual and augmented reality. These disparate stories, woven throughout different race series and generations, all entwine into a tapestry that shows what's to come.

It's a future we should be excited about. And it's a fitting place to end my journey into the world of motorsport science.

This book has been a celebration of creativity. Whether it's in engines, brakes, aerodynamics, safety, programming, materials, manufacture, tyres or fuels, motorsport's reach extends far beyond the track and into every corner of our lives. It reveals the invisible threads that bind us together. It's a demonstration that all knowledge, no matter how seemingly unrelated, is linked. There is no such thing as an isolated discovery any more. We live in a time when a car's airflow duct can build better skyscrapers and a pencil shaving on a piece of tape can fast-charge our mobile phone. Those who sneer at racing, slamming it as a wasteful and frivolous activity, are blind to these incredible advances.

Of course, there are still naysayers about many aspects of the green technologies we've covered. People who laugh at the idea of electric vehicles and are unable to imagine a world without the fire and fury of an ICE. They're the same people who thought that seat belts were a bad idea. Don't listen to them. Watch the race and see more.

Motorsport excels in demonstrating ideas and proving doubters wrong. It's a tradition that stretches back through time, past David Coulthard and his Xbox graphics card, beyond the heyday of Sir Jackie Stewart and his campaigns for basic safety, right to the birth of racing cars themselves, with Camille Jenatzy pushing the edges of speed. Humans are storytellers, and we are captivated by those who live their lives one ten-thousandth of a second at a time. Racing ignites a deep, lasting passion in its fans. And that means it doesn't just have a role in coming up with green ideas; it has a duty to popularise them, too.

Around 10 per cent of the world's population watches Formula One. The technologies they see on track show what is possible. And while some changes meet initial resistance, this never lasts for long. Many of the sport's stars also use their leverage to promote green technologies; every year, Rosberg hosts the GreenTech Festival, placing sustainability at the heart of his agenda. Others, such as Hamilton, focus on diversity and inclusivity. These steps chip away at larger societal problems and make life a little better.

Formula E was, of course, established to promote new, greener technologies. In 2015, a report by Ernst & Young found that, over 25 years, Formula E could help sell 52–77 million electric vehicles, with a resultant saving of 900 million tonnes of CO_2 emissions. Combined with Extreme E, this number could be more than 1 billion tonnes saved. Everyone knows the importance of its legacy. 'The reason you become a sportsman is to be remembered as the best in your sport,' one of its champions, Jean-Éric Vergne, told an online session in 2020. 'But with Formula E, you can leave a legacy for the next generation – not just remembered as the best but as the one who inspired their fans to make a change.' Vergne, just like other FE drivers such as Lucas di Grassi and Sam Bird, knows more than pride is at stake when he enters the cockpit.

Above them both, the FIA has numerous initiatives to promote safer driving, traffic awareness and better motoring laws around the world. A prime example is its 3,500 Lives campaign, based on the number of people still killed on our planet's roads every day. The FIA is also the primary sponsor of the International Road Assessment Programme, a rating system aiming to take the most dangerous roads in the world – such as those lacking safe places to turn – and improve them.

If it achieves its goal of making all roads 3-star or better, it's estimated it could help saves 467,000 lives a year and prevent 100 million serious incidents over the next two decades.

Even NASCAR, seemingly the black sheep of sport's sustainability revolution, is playing its part. In addition to recycling its tyres and engine oil, it has one track, Pocono Raceway, powered completely by solar energy, and has encouraged its fans to recycle more than 25 million bottles and cans at its race events. While these may be small acts in a much larger problem, NASCAR's ability to speak to millions of die-hard fans across the US can only have a beneficial effect in raising awareness of why environmentalism matters. And, ultimately, stock car racing is driven by manufacturers. When the world goes electric within the next decade, you can be sure NASCAR will too.

In Ushuaia, a sharp wind cuts along the jetty. I watch the sea birds on the beach, listen to the gentle lap of the tide dancing on the pebble-strewn shore and rest under the majesty of peaks that soar up to touch the deep-blue Southern sky. It's as silent as my journey's start, in those hushed halls of Maranello, among the red-liveried triumphs of the Prancing Horse.

It feels like I've come a long way, and that I've still only just begun. All the technologies and ideas in this book are in their infancy. Some will succeed and some will fail. They are a vanguard of a new era.

Somewhere, lost in the lilac Patagonian mists, an engine starts. Its meaty roar echoes around the serene bay, breaking the spell. I take one last lungful of pure Antarctic air and head back to the ship, thinking of a young, eager scientist I once met who was looking to get into motor racing. He was in his early 20s, part of his university's Formula SAE team, thumbs blackened with grease and eyes sparkling with the look of someone who'd caught the bug. He wanted to win, he told me, of course he did. But, win or lose, he'd go again. It's such spirit that drives people to do incredible things.

'In racing,' he said, 'you get to try things that you'd never try elsewhere. And sure, sometimes they don't work. But when they do? When they do, it's *awesome.*'

He was right. When they work, they change the world.

Appendix: Martin, Mosler & Me

During interviews for my first book, the most common question asked was why I decide to tackle such unusual science stories. The tale of how *Racing Green* came about is worth telling, but is so unbelievably weird it doesn't fit anywhere in the book proper. It involves a Le Mans racer, a spaceship covered in cat poo and an eccentric millionaire live from his remote Caribbean island.

⋆ ⋆ ⋆

If you'd told me I'd write a book about sports 20 years ago, I'd have laughed at you. Back then, I was an overweight teen, bullied at school and with a healthy loathing for, not to mention distrust of, competitive games. On my first day at school, I'd been lined up against a wall with the rest of my year and divided by eye into 'kids who can run fast' and 'kids who are worthless'. I was in the worthless camp. For the next few years, I received no coaching, training or encouragement but plenty of abuse from PE teachers. Any love for sport was burnt out of my soul and, rather than appreciate the nuance, dedication and skill of modern athletes, I came to see them only as assholes who probably tormented weaker kids to get where they were. Despite a brief flirtation with rugby (hey, we all experiment at university), the only sport I followed was motor racing.

I didn't exercise for close to 15 years, getting chubbier with each passing day. Eventually, I decided I didn't need to have three chins and hired a personal trainer to help me lose a few hundred pounds. It was here that fate struck. One day, the trainer couldn't make it to the fitness centre and suggested I come by another client's home – with his permission – for a session in his private gym. Heading off after an hour of sweating, straining and swearing, I came face-to-face with the homeowner chatting to a friend on his driveway. Figuring it was only polite to say thanks for letting me perspire all over his rowing machine, I walked over and introduced myself.

The house belonged to Martin Short. The friend was David Brabham. And behind them was Brabham's BT62 – a supercar valued at a cool £1.2 million.

Rather than tell me to bugger off, Short was a gracious host. A former RAF veteran turned professional race team owner and driver, in 2003 he'd decided (with help from a sponsor) to buy two V10 Dallara SP1 Prototypes and enter the 24 Hours of Le Mans. In his first year, he was running fourth until a shunt from Sébastien Bourdais caused a mechanical fault and resulted in him crashing off track spectacularly. The next year, Short was running well when he noticed helicopters circling overhead and figured he was somewhere on track close to the leader. Soon after, he saw an Audi loom large in his mirrors and, assuming he was being lapped, let it pass. It was only when he returned to the pits that Short discovered *he* had been leading the most prestigious race in motorsport and had let the second-placed Audi slip through uncontested. When you look at the list of Eric Thompson Trophy winners – presented to the highest-placed British driver at Le Mans each year – the usual names of Allan McNish, Johnny Herbert and Anthony Davidson are rudely interrupted in 2007 by those of Martin Short and Stuart Hall. A tiny outfit from rural Cambridgeshire had, for the briefest of moments, outpaced the greatest car manufacturers in the world.

I liked Short immediately; he was a raconteur and he had the stories to back it up. Soon we were gossiping about the finer, dorkier points of the Brabham's design and I was enthralled by yarns of chasing petrol-fuelled glory. Then, with a gleam in his eye, Short beckoned me into his garage. 'If you think that Brabham's cool,' he said, 'you're going to love this.'

My heart did but my eyes didn't. In the garage was, without question, the ugliest car I've ever seen. It looked like an upside-down sled, the kind of hollowed-out plastic bathtub kids sit on to slip down a hill for a few metres before bouncing out of control. Rising out of this odd design, nose curving down and arse sticking flat out, was a cab that poked up near-vertically before tapering away. The wheels were flush inside the box body and it was painted a battleship silvery grey. If the exterior felt like a spare prop from a dodgy sci-fi B-movie, inside the car wasn't much better. The fit and finish was dusty and basic, with far too many dials and gauges on the dashboard, and there was a conspicuous curl of cat poop on the passenger seat from where an errant moggy had crawled through an open window.

The car was a Mosler Consulier GTP. Back in the late 1980s and early 1990s, it was pretty much unbeatable.

A week later, I found myself on a video call with the car's creator. Warren Mosler made his millions as a hedge fund manager on Wall Street before becoming a leading proponent of Modern Monetary Theory – essentially that money as we know it is a government monopoly – and Mosler's Law, that no financial crisis can't be solved by changing up fiscal policy. (Not being an economist, I am pretty sure I've messed up either or both of those definitions.) In addition to economics, Mosler has run for public office numerous times, has run rings around pundits on US TV and is a visiting professor at several universities around the world. But his real passion is for cars. After his success on Wall Street, Mosler got into racing as a hobby, crunched the numbers and realised that if he could build a lightweight supercar, it could dominate anything on track. The result was the Consulier.

Built out of a carbon fibre–Kevlar composite and weighing a third less than its rivals, the Consulier was 40 years ahead of its time. The reason for its lack of looks is simple: it's a car based on numbers, with virtually no consideration given to aesthetics. If a design element made the car light and fast, it was included. If it was just there to look pretty, it was absent. The Consulier wasn't created to win a beauty pageant; it was created to leave its rivals in the dust.

And so it proved. The Consulier won a string of International Motorsports Association (IMSA) and 24-hour races, as well as *Car and Driver*'s One Lap of America. Mosler even put up $100,000, challenging anyone to beat his car on a single lap. No one could.

Eventually, tired of the upstart company dominating the grid, the IMSA started giving the Consulier a 136kg weight penalty. After it still kept winning, the car was banned outright in 1991. A subsequent Mosler design, the more eye-catching MT900, was even more potent. There was talk about it entering another major race series when suddenly a rule was added, requiring a certain number of the cars be sold before they could be considered for the competition. 'The joke going around at the time,' Short recalls, 'was that the number was whatever Mosler managed to sell, plus one.'

With racing no longer an option, Mosler Automotive had to rely on general sales. Unfortunately, there weren't many. Although on paper a Mosler rivalled anything on the road, it didn't have a marque like Ferrari, Lamborghini or Porsche. Nor did it have the pointless luxury trappings that slow a car down but make great photos on social media. Mosler's cars were fast. Mosler's cars were revolutionary. And Mosler's cars were flops.

I chatted with Mosler about the Consulier. He schooled me on economic theory and we laughed about his time visiting Ferrari's wind tunnel at Maranello. I've had the privilege in my career of interviewing about a dozen Nobel prize winners and Mosler spoke with the same arrow-like clarity they've all possessed, a direct and incisive way of cutting out dross. He struck me as a brilliant mind who saw the world in pure figures, rather than existing in the space among heartbeats where racers live or die. There seemed to be a story to tell. But it was only when he told me about his ferry that this book started to take form in my mind.

Mosler now lives on Saint Croix in the US Virgin Islands, a former pirate haven south-east of Puerto Rico. Its nearest neighbour is Saint Thomas, 42 miles away to the north, separated by a choppy channel liable to make all but the heartiest passengers a little nauseous. Mosler decided to design and build a ferry link between the islands. As you might expect, it isn't a conventional boat. Instead, it's a catamaran, with a long, thin passenger cabin resting atop four small hulls. Not only is it lighter, faster and more fuel efficient than any of its rivals, the unusual four-hulled design (called a 'tandem cat') has an added bonus: it stabilises the passenger cabin, creating a smoother ride and preventing seasickness.

The tandem cat captured my imagination. I sat down and began scribbling a list of different success stories I knew that had emerged from motor racing. Rear-view mirrors, crash helmets, the HANS device, electric cars, neonatal transport, tyre recycling … I sent a quickfire email to my publisher. I had an idea for a book about technology coming out of motorsport. And not just any technology, either – green advances that improve our lives in ways you'd never think had anything to do with cars.

This book is the result.

Acknowledgements

The famous curse 'May you live in interesting times' comes to mind when I think about writing this book. Work began in late 2019, and I expected to have completed all of the interviews and investigations within the space of six months. Instead, a pandemic swept the world and staying at home became necessary to save lives. Plans were made and plans were changed. One visit was rescheduled eight times before, eventually, falling through. In total, this book was written during three separate lockdowns and a two-week quarantine in a South Korean hotel room while I somehow simultaneously managed to start and finish a PhD. It wasn't a straightforward assignment.

All of which makes it important to thank the people who made this book possible. Some, I have known for a long time. Sheila Chapman and Allison Holloway once again read pages and gave advice; while this book wouldn't exist if I hadn't been working out with Marco Galea in someone else's house. New friends, too, have aided me on this journey. Martin Short gave time, made introductions and welcomed me into his family world; Jan Zabkiewicz and Stephen Docherty brought petrolhead passion to a hasty sense-check; and the regulars of the History Hack podcast gave me moral support when I needed it most. Thanks are also owed to the crew of the *Queen Victoria*, who saw me safe across the Atlantic (twice), down the Amazon river and around Cape Horn, all during the greatest crisis of the twenty-first century.

Part of the fun in writing this book was covering so many areas, but that means I had to trust the experts where my knowledge fell down. This includes all interviewed in the book (none of whom knew me, all of whom gave full cooperation); as well as input from Claire Benson at London South Bank University on fire; Kristy Turner at the University of Manchester on organic chemistry; and Fernando Gomollón Bel for arranging an interview on graphene. Special thanks must be given to Envision Virgin Racing, Extreme E and Venturi, none of whom hesitated for a moment when I told them my plans and asked for press access into their secretive worlds. In particular, my thanks to Sam Bird, who put up with stupid questions from my weary, jetlagged brain with grace and charm. It's no surprise to me he's won a Formula E race in every season of the sport; if you can put up with me, you can keep your cool under any level of pressure.

Finally, a heartfelt apology to the student team at the University of Alberta, Edmonton, and to Delft's Maxime van Kekem, whose contributions didn't make the final cut. You are the future of science and engineering, and your kindness is not forgotten.

Dr Kit Chapman

Index